Springer Theses

Recognizing Outstanding Ph.D. Research

For further volumes:
http://www.springer.com/series/8790

Aims and Scope

The series "Springer Theses" brings together a selection of the very best Ph.D. theses from around the world and across the physical sciences. Nominated and endorsed by two recognized specialists, each published volume has been selected for its scientific excellence and the high impact of its contents for the pertinent field of research. For greater accessibility to non-specialists, the published versions include an extended introduction, as well as a foreword by the student's supervisor explaining the special relevance of the work for the field. As a whole, the series will provide a valuable resource both for newcomers to the research fields described, and for other scientists seeking detailed background information on special questions. Finally, it provides an accredited documentation of the valuable contributions made by today's younger generation of scientists.

Theses are accepted into the series by invited nomination only and must fulfill all of the following criteria

- They must be written in good English.
- The topic should fall within the confines of Chemistry, Physics, Earth Sciences, Engineering and related interdisciplinary fields such as Materials, Nanoscience, Chemical Engineering, Complex Systems and Biophysics.
- The work reported in the thesis must represent a significant scientific advance.
- If the thesis includes previously published material, permission to reproduce this must be gained from the respective copyright holder.
- They must have been examined and passed during the 12 months prior to nomination.
- Each thesis should include a foreword by the supervisor outlining the significance of its content.
- The theses should have a clearly defined structure including an introduction accessible to scientists not expert in that particular field.

Margarita Puentes Vargas

Planar Metamaterial Based Microwave Sensor Arrays for Biomedical Analysis and Treatment

Doctoral Thesis accepted by
Technische Universität Darmstadt,
Darmstadt, Germany

Springer

Author
Dr. Margarita Puentes Vargas
Department of Electrical Engineering and Information Technology
Institute of Microwave Engineering and Photonics
Technische Universität Darmstadt
Darmstadt
Germany

Supervisor
Prof. Dr.-Ing. Rolf Jakoby
Department of Electrical Engineering and Information Technology
Institute of Microwave Engineering and Photonics
Technische Universität Darmstadt
Darmstadt
Germany

ISSN 2190-5053 ISSN 2190-5061 (electronic)
ISBN 978-3-319-06040-8 ISBN 978-3-319-06041-5 (eBook)
DOI 10.1007/978-3-319-06041-5
Springer Cham Heidelberg New York Dordrecht London

Library of Congress Control Number: 2014935055

Printed on acid-free paper

Springer is part of Springer Science+Business Media (www.springer.com)

Parts of this thesis have been published in the following documents:

Journals

M. Puentes, M. Schüßler, C. Damm and R. Jakoby, "Extraction of Capacitive Profiles with CRLH Transmission Lines". *Frequenz Magazine*. 5/6-2010.

M. Puentes, M. Schüßler, C. Damm and R. Jakoby, "Extraction of Capacitive Profiles with a Planar Metamaterial Sensor". *Journal of Applied Physics*. Volume 103, Issue 3, pp 815–819, June 2011.

M. Puentes, M. Maasch, M. Schüßler and R. Jakoby, "Frequency multiplexed 2-dimensional sensor array based on split-ring resonators for organic tissue analysis". *IEEE Transactions on Microwave Theory and Techniques*, Vol 60, Issue 6, pp 1720–1727, June 2012.

M. Schüßler, C. Mandel, M. Puentes and R. Jakoby, "Metamaterial Inspired Microwave Sensors". *Microwave Magazine*, March–April 2012.

International Conferences and Workshops

M. Puentes, B. Stelling, M. Schüßler, A. Penirschke and R. Jakoby, "Planar Sensor for Permittivity and Velocity Detection Based on Metamaterial Transmission Line Resonator". *European Microwave Conference*, Roma, Italy, September 2009.

M. Puentes, B. Stelling, M. Schüßler, A. Penirschke, C. Damm and R. Jakoby, "Dual Mode Sensor for Belt Conveyor Systems Based on Planar Metamaterials", *IEEE Sensors Conference*, Christchurch, New Zealand, October 2009.

M. Puentes, M. Schüßler, A. Penirschke, C. Damm and R. Jakoby, "Metamaterials in Microwave Sensing Applications", *IEEE Sensors Conference*. Hawaii, USA, November 2010.

M. Schüßler, M. Puentes, C. Mandel and R. Jakoby, "Multi-Resonant Perturbation Method for Capacitive Sensing with Composite Right-Left Handed Transmission Line". *International Microwave Symposium*, Anaheim, USA, June 2010.

M. Puentes, M. Schüßler and R. Jakoby, "Sensor Array Based on Split Ring Resonators for Analysis of Organic Tissues". *International Microwave Symposium*, Baltimore, USA, June 2011.

M. Puentes, M. Schüßler, and R. Jakoby, "2D Sensor Array Based on Split Rings Resonators for Monitoring of Organic Tissue". *IEEE Sensors Conference*, Limerick, Ireland, October 2011.

M. Puentes, F. Bashir, M. Schüßler and R. Jakoby, "Dual Mode Microwave Tool for Dielectric Analysis and Thermal Ablation Treatment of Organic Tissue". *International Conference of the Engineering in Medicine and Biology Society*, San Diego, USA, August 2012.

M. Schüßler, M. Puentes, D. Dubuc, K. Grenier and R. Jakoby, "Simultaneous Dielectric Monitoring of Micro Fluidic Channels at Microwaves Utilizing a Metamaterial Transmission Line Structure". *International Conference of the Engineering in Medicine and Biology Society*, San Diego, USA, August 2012.

M. Puentes, F. Bashir, M. Maasch, M. Schüßler and R. Jakoby, "Planar Microwave Sensor for Thermal Ablation of Organic Tissue". *European Microwave Conference*, Nuremberg, Germany, October 2013.

Supervisor's Foreword

It is a great pleasure to introduce Dr. Margarita Puentes's thesis work, accepted for publication within Springer Theses and awarded with a prize for an outstanding original work. Dr. Puentes joined my research group for Microwave Engineering at Technische Universität Darmstadt in August 2008. She started her doctoral study with a nine-month scholarship of the DFG-Graduiertenkolleg 1037 "Tunable Integrated Components in Microwave Technology and Optics" and continued her studies within a full position as a Research Assistant. She completed it with an oral defense on November 2013. Dr. Puentes's thesis includes significant original scientific contributions, representing considerable advancement in the field of biomedical microwave sensing, which she published as first author in top journals and international well-recognized conferences.

Microwave sensors, besides radio detection and ranging, offer the opportunity of contactless (non-invasive) or minimal-invasive sensing and benefit from different material properties at high frequencies as well as from the electromagnetic interaction between the sensor structures and the material under test, offering the potential for higher sensitivity. For biomedical applications, the major objective of the thesis work of Margarita Puentes, was a systematic analysis of planar microwave sensor arrays to derive the dielectric properties via the frequency characteristics of biological cells and organic tissue. As sensing elements she used Composite Right/Left-Handed (CRLH) transmission lines and Split-Ring Resonator (SRR) unit cells, with dimensions well below the operating wavelength, detuned by dielectric loading, where the resulting shift of a distinct resonance frequency of one unit cell or various resonance frequencies of an array of unit cells is being measured and analyzed.

Her newfound way to use sensitive periodic structures in combination with robust extraction techniques give rise to sensor arrays that can monitor several material samples simultaneously with one readout signal or the sample's dielectric profile or inhomogeneity's. This can be done with high flexibility in the operating frequency and geometry, allowing to adapt the sensor's shape and size, depending on the various requirements for material characterization, e.g., to bring the sensor elements as close as possible to the material under test. Most impressive, beside novel approaches, concepts and designs, is the proof-of-concept by many demonstrators and her creativity to enter this interdisciplinary biomedical field with focus on: (1) Biological screening using microfluidic technology, to test several samples of cells simultaneously, aiming in a long run for lab-on-chip devices. This work has been done in collaboration with CNRS-LAAS Institute in Toulouse, France. (2) Minimal invasive screening of organic tissue for abnormalities such as cancerous tissue, and at the same time, to treat the found abnormalities by means of thermal ablation therapy. The dual mode feature of these devices has been done for the first time to my knowledge. In the detection mode, with input power of few mW, Dr. Puentes uses different sensor structures based on SRRs at different operating frequencies to obtain the relative dielectric image of the tissue. On the other hand, the same structures offer the second mode of operation to perform thermal ablation treatment in a highly focused way with a considerable low input power of up to 10 W, where measurements are carried out with adapted phantoms of organic tissue and the sensors are able to heat a significant ablation zone from 40 to 110 °C. Introducing this approach of sensing and treatment with the same microwave sensor device offers several benefits such as the automatic alignment between the determined tumor location and subsequent heat application as well as the control of the ablated area by using the tool in the sensing mode after the ablation procedure is completed. The next step is to prove this concept on humans in collaboration with scientists in the medical field.

Darmstadt, February 2014 Rolf Jakoby

Acknowledgments

The thesis presented here is the result of my work as a scholarship student and afterwards research assistant in the Institute for Microwave and Photonics at the Technische Universität Darmstadt. Apart from my efforts, the success of any project depends largely on the encouragement and guidance of others. For these reasons, I would like to thank all the persons involved in the development of this work.

First of all, I want to thank Prof. Dr.-Ing Rolf Jakoby for giving me the opportunity to pursue my doctoral studies. He was always supportive and allowed me to direct my research into my desired topic. Finally, he allowed me the flexibility required to combine my work with my personal situation. I would also like to thank Prof. Dr. Ferran Martín for being the second examiner of my work and all his important contributions to it.

A special recognition has to be given to Dr.-Ing Martin Schüßler. Since my arrival to Darmstadt, he was highly understanding, supportive, and made substantial contribution to the adequate development of my work. I want to thank him specially for his patience, friendship, and helping me feel more at home.

To all my colleagues I would like to thank them for receiving me into their group and giving the necessary support to overcome the obstacles encountered. Every time I was in need of support I found it without difficulty. Special thanks to Dr.-Ing Christian Damm, Matthias Maasch and Dr.-Ing Holger Maune for their advice in all relevant scientific discussions and experimental development. I would like also to mention Peter Kieslich and Andreas Semrad for their technical support in the development of my sensors.

To my family and friends, thank you for believing in me and always encouraging me to continue pursuing my goals. As for my parents who inspired and motivated me, and specially my mother, thank you for giving me the education necessary to arrive to the point where I am today. Thank you for encouraging me to go abroad to continue my career although it meant a long separation, it was the right decision and I am thankful for the opportunity.

And finally to my wonderful and smart daughter Sofia, thank you for joining me in this experience and giving me the time to develop my work. Thank you for being a very understanding and easygoing child that not only accompanied me many times in the day to day work activities but also found a way to enjoy it and integrate herself within the group of colleagues.

Darmstadt, October 2013 Margarita Puentes Vargas

Contents

Symbols and Abbreviations

α	Attenuation constant
β	Phase propagation constant
B	Magnetic flux
BW	Bandwidth
C_{RH}	Right-handed capacitor
C_{LH}	Left-handed capacitor
dz	Infinitesimal unit cell length
CCF I	Cauer canonical form I
CCF II	Cauer canonical form II
CHO	Chinese Hamster Ovary cells
CPW	Coplanar waveguide
CRLH-TL	Composite right-left handed transmission line
CT	Computerized tomography scans
D	Electric flux
Δz	Finite unit cell length
E	Electric field strength
ε_0	Free space permittivity
ε_r	Relative permittivity
η	Wave impedance
f	Frequency
f_s	Series frequency
f_p	Parallel frequency
FoM	Figure of merit
γ	Propagation constant
H	Magnetic field strength
j, i	Imaginary unit
λ	Wavelength
ℓ	Length
L_{RH}	Right-handed inductor
L_{LH}	Left-handed inductor
LHM	Left-handed materials
μ_0	Free space permeability
μ_r	Relative permeability
MCEM	Mixed Cauer expansion method

MRI	Magnetic resonance imaging
MRPM	Multi-resonant perturbation method
MUT	Material under test
MWA	Microwave ablation
n	Number of unit cells
NaCl	Salt
ω	Angular frequency
ω_0	Angular transition frequency
ω_s	Angular series cut-off frequency
ω_p	Angular parallel cut-off frequency
PDMS	Polydimethylsiloxane
PLH-TL, LH-TL	Pure left-handed transmission line
PRH-TL, RH-TL	Pure right-handed transmission line
RFA	Radio-frequency ablation
RHM	Right-handed materials
RPM	Resonant perturbation method
RT/D	Rogers Duroid substrate
RTD	Resistance temperature devices
S_{11}	Input reflection coefficient
S_{12}	Backward transmission coefficient
S_{21}	Forward transmission coefficient
S_{22}	Output reflection coefficient
SAR	Specific absorbtion rate
SRR	Split ring resonator
SMD	Surface mounted devices
τ_d	Group delay
TDR	Time domain reflectometry
TUT	Tissue under test
v_p	Phase velocity
v_g	Group velocity
Y	Complex admittance
Z	Complex impedance
Z_B	Bloch impedance
Z_{in}	Input impedance
Z_L	Line impedance
Z_0	Reference impedance

Chapter 1
Introduction

The modern industry is interested in new universal adaptable and affordable sensor concepts in a wide range of applications which can be satisfied with microwave sensors. They are very versatile and can perform in different environments ranging from industrial, environmental to even biomedical applications. In many scenarios where dielectric properties of a certain object are of importance, microwave sensors can be employed. Their advantages lie in the facts that they can sense in a non-invasive and contact-less way. They can achieve a deeper penetration, compared to terahertz or optical techniques, with the drawback of lower resolution. Another reason is that most materials, organic and inorganic, have a specific frequency behavior which can provide information about their structure and function. Microwave sensor arrays are able to extract this information by retrieving these spectral signatures and give a spatial distribution of several materials under test (MUTs) simultaneously.

A very interesting approach is the design and development of microwave sensors based on metamaterial structures. They are artificial materials made of a periodic arrangement of unit cells much smaller than the guided wavelength. They exhibit properties not available in nature, such as simultaneous negative permittivity and permeability. For applications where propagating electromagnetic fields are used, metamaterials display interesting effects such as backward wave propagation due to an antiparallel group and phase velocity resulting in the backward travel of the wavefronts while the energy still travels forward away from the source to satisfy causality. Metamaterials can be used for several interesting devices in microwave and terahertz engineering such as antennas, sensors, matching networks among others. The development of metamaterial based sensors offers several advantages. First, under certain design conditions, their size can be much smaller than conventional structures making them suitable for miniaturized scenarios that are desired in a wide range of applications. They are also very flexible in terms of operating frequency and geometry of the structures, they can be developed from cylindrical shapes to planar geometries without loosing any of their properties. Additionally, they are perfect for schemes where the aim is to extract a spatially resolved profile of the sensed object.

M. Puentes Vargas, *Planar Metamaterial Based Microwave Sensor Arrays for Biomedical Analysis and Treatment*, Springer Theses,
DOI: 10.1007/978-3-319-06041-5_1,

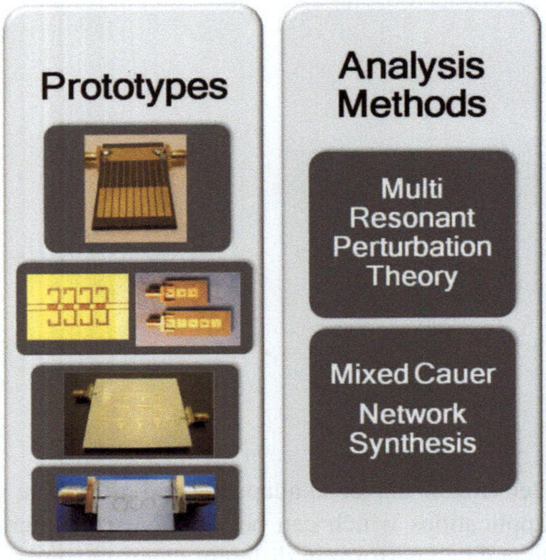

Fig. 1.1 Graphic overview of the planar sensors, extraction techniques and studied applications presented in this thesis

In this work a novel approach for microwave planar sensor arrays using periodic structures as building elements is investigated. One of the main novelties of the proposed planar sensor concept is the array feature which allows to make a simultaneous analysis of several materials under test with a single readout signal, or a relative characterization of one material with information about its spatial distribution. Additionally, the use of metamaterial structures as building elements for the development of the sensors gives them high flexibility in terms of geometry and operating frequency, enabling miniaturization and increased sensitivity by enhanced field interaction. The advantage of subwavelength image pixel resolution resulting from the discrete metamaterial unit cells was found as an additional feature. The transmission line approach to design metamaterial structures was used, where a unit cell made of inductors and capacitors is periodically cascaded. For one specific scenario, namely for analysis and treatment of organic tissues, the resonant approach is used. The reason is that the sensors presented for this application operate in two different modes, sensing and treatment of organic tissues and in the second case a clear resonance frequency is advantageous. An overview of the work done in this thesis is depicted in Fig. 1.1. The first column includes several of the developed prototypes of planar microwave sensor arrays, in the middle column the two main extraction techniques used for obtaining the dielectric image of a specific MUT are shown and finally, the right column presents the chosen application scenarios to prove the wide functionality and high usability of the developed devices. All of the sensors share as principle of operation that they are designed for direct contact with the MUT. They enable the application of a probe signal allowing the electric field to interact efficiently with the MUT depending on the specific design of the sensor and the dielectric properties of the MUT. Evaluation of the reflected and/or transmitted signal delivers relative dielectric properties and material distribution of inhomogeneous materials in the vicinity of the sensor through the extraction of a capacitive/dielectric profile.

Two extraction techniques are investigated in order to derive the dielectric properties from the measured sensor signals. First, the multi resonant perturbation method (MRPM) is used [1], where the equivalent circuit of the sensor structures is the Composite Right/Left-handed transmission line (CRLH-TL). The method is able to extract the left-handed capacitive profile $C_{LH,n}$ of a periodic structure influenced by a MUT. The second method is completely developed within the framework of this work and it is based on network synthesis. It is called the mixed Cauer expansion method (MCEM) and it is able to extract the complete equivalent circuit of the periodic structure as long as it can also be modeled as a CRLH-TL. The MRPM and the MCEM are comparable in terms of what they can achieve. Both methods are able to extract a dielectric image of the MUT with a one unit cell resolution, and therefore, the position of abnormalities can also be located. An extensive comparison with the advantages and disadvantages of the methods will be given within this thesis [2].

The proposed sensor basic concept is investigated and analyzed in terms of why the use of a periodic structure to create an array type sensor is advantageous. Also the combination with the mentioned extraction techniques is thoroughly tested. An initial assessment of the developed sensor concept is done for monitoring the quality of paper. The reason to test the sensors with this particular material under test, is due to the fact that in industrial environments, the sensing of products throughout the manufacture process is highly important to control the quality of the final product. Microwave sensors are well suited for this purpose. Although these sensors are generally more expensive than sensors working at lower frequencies, they feature properties that make the cost worthwhile. Specially for moving environments, which can be found in many industrial sites, microwave sensors are advantageous, since they provide contact-less and therefore non-invasive sensing techniques. The solution that will be presented includes sensors based on metamaterial structures that contain 3 to 9 unit cells. One of the presented sensors exhibits a unique planar geometry where the complete structure is located on the surface of the sensor enabling high Q inductors with large values. The other main design is based on a traditional coplanar waveguide. By using the extraction techniques, different dielectric profiles are adequately extracted with the proposed structures for both, dry and moist, paper samples. The sensors are able to detect the properties of the MUT at a distance of up to 2.5 cm and changes in the capacitance of approximately 0.1 %.

To explore the wide use of possible applications for the presented sensor concept, two application scenarios are studied. The first scenario of application is for the frequency characterization of biological cells for cytological screening. The traditional way to analyze pools of cells is done by looking at the sample under the microscope and using markers or fluorescent agents that can alter the biological activity of the MUT. When a large amount of cells needs to be analyzed, the process is long and cumbersome. The goal is to develop a microwave sensor array system by combining planar periodic metamaterial structures with microfluidic channel technology. Thus, the analysis of the biological cells with this sensor array, e.g. in order to extract properties for diagnostic purposes, can be performed almost automated by measuring the dielectric properties, and hence, reduces significantly the process time. The sensors are made on a coplanar waveguide design and have 1–4 unit cells. The final proof of concept is given when different concentrations of chinese hamster ovary (CHO) cells, a well-known mammalian cell model for biological and medical research, are successfully extracted [3].

The second application scenario is in the medical field. In this case, microwave sensors are used to analyze dielectric properties of organic tissue for the detection of abnormalities such as cancerous cells [4]. Furthermore, the tool can offer a course of treatment for eradicating the abnormalities by means of thermal ablation. For the devices that will be presented, split ring resonators (SRRs) were used as resonant particles coupled to microstrip lines. The principle of operation includes the sensor coming into direct contact with the tissue through a minimal invasive procedure and extracting the relative dielectric image. Since the goal is to use these structures as part of a minimal invasive tool their size and miniaturization possibility are very important and can be obtained using metamaterial inspired structures. The possibility to have a dielectric image with several pixels of resolution will also be addressed and was another reason for choosing metamaterial structures in the design. Once an abnormality is detected by the sensor array, its exact position is known and the device can operate in a way that the maximum field is concentrated on one specific SRR for the treatment mode. As a result, the temperature of the tissue surrounding it increases to a level adequate to perform thermal ablation. The devices presented here range between 1 and 12 SRRs, or better called pixels, and are able to heat the tissue to a temperature of approximately 109 °C with an input power of 10 W [5, 6].

The novelty of the work presented in this thesis lies in the newfound way to use periodic structures in combination with robust extraction techniques to create sensor arrays that can monitor several samples of MUT simultaneously to perform not only material characterization but also obtain the spatial distribution of the dielectric information. The sensors have the advantage of high flexibility in the geometry and frequency of operation making them suitable for a wide range of application environments. Furthermore, they can be tailored to perform not only non-invasive sensing but also actively interact with the MUT in the case of the thermal ablation treatment.

1.1 Outline

The outline of this book is as follows. Chapter 2 introduces the fundamentals of metamaterial structures where the transmission line approach and the differences between the infinitesimal unit cells and the real finite length unit cells are highlighted. Additionally, the resonant approach is addressed specifically using microstrip coupled SRRs.

Chapter 3 includes the extraction techniques used to detect dielectric profiles of material under test. First, the multi resonant perturbation method is explained, then the mixed Cauer network synthesis is introduced and studied. Finally, a comparison of the two methods is presented.

Chapter 4 includes the investigations toward the novel sensor basic concept and includes the first planar microwave sensors designed within the framework of this work. The advantages of using periodic structures are also addressed. An initial proof of concept is performed with the designed structures by monitoring different types of paper, some with high moisture content. Different dielectric profiles of abnormalities in the paper were successfully extracted with the extraction methods studied in the previous chapter.

Chapter 5 presents the first application scenario where the planar microwave sensor arrays are used for a biomedical application. The prototypes were developed within a close collaboration with the CNRS-LAAS institute in Toulouse, France. The prototypes are integrated with microfluidic technology and are able to extract different concentrations of Chinese hamster ovary (CHO) cells. They show great promise to be part of a lab on chip application.

Chapter 6 addresses the final area of application in the medical field. Different microwave planar sensors based on SRRs will be presented. They include optimized designs to perform the sensing of the dielectric properties of different organic tissues, and furthermore, perform treatment of cancerous tissues with the same structures by means of thermal ablation.

Chapter 7 presents the conclusions of the presented work and an outlook to possible future developments.

References

1. M. Schüßler, M. Puentes, C. Mandel, R. Jakoby, Multi-resonant perturbation method for capacitive sensing with composite right-left handed transmission line. in *Proceedings of IEEE MTT-S International Microwave Symposium Digest* (Anaheim, USA, 2010)
2. M. Puentes, M. Schüßler, C. Damm, R. Jakoby, Extraction of capacitive profiles with CRLH transmission lines. Frequenz **5/6**, 74–78 (2010)
3. M. Schüßler, M. Puentes, D. Dubuc, K. Grenier, R. Jakoby. Simultaneous dielectric monitoring of micro fluidic channels at microwaves utilizing a metamaterial transmission line structure. in *Proceedings of 34th Annual International Conference of the Engineering in Medicine and Biology Society* (San Diego, USA, 2012)
4. M. Puentes, C. Weiß, M. Schüßler, R. Jakoby, Sensor array based on split ring resonators for analysis of organic tissues. in *Proceedings of IEEE MTT-S International Microwave Symposium Digest* (Baltimore, USA, 2011)
5. M. Puentes, F. Bashir, M. Schüßler, R. Jakoby, Dual mode microwave tool for dielectric analysis and thermal ablation treatment of organic tissue. in *Proceedings of 34th Annual International Conference of the Engineering in Medicine and Biology Society* (San Diego, USA, 2012)
6. M. Puentes, M. Maasch, M. Schüßler, R. Jakoby, Frequency multiplexed 2-dimensional sensor array based on split-ring resonators for organic tissue analysis. IEEE Trans. Microwave Theor. Tech. **60**(6), 1720–1727 (2012)

Chapter 2
Fundamentals of Metamaterial Structures

2.1 Introduction to Metamaterials

In the last decade a new research area based on the study of Metamaterials has emerged. There is not an absolute definition of Metamaterials but the most commonly used is that they are artificial materials made of a periodic arrangement of unit cells much smaller than the wavelength with some unique properties [1–3]. Such properties include a simultaneous negative permittivity and permeability that translates into a negative refractive index, these structures are commonly referred as left-handed materials (LHM). By using LHM in the design of a structure, its properties can be tailored in a broader manner than with classical materials. Interesting effects can be observed in the propagation of the electromagnetic waves, for example, backward propagation of waves due to an antiparallel group and phase velocity resulting in the backward travel of the wavefronts while the energy still travels forward away from the source to satisfy causality. Another desired effect is the significant reduction in size compared to classical structures that can be achieved for certain conditions relevant to a specific design. Metamaterials can be used for several interesting devices in microwave and terahertz engineering such as antennas, sensors, matching networks or new superlens with a higher focusing capability than conventional ones in areas such as telecommunications and biomedical applications.

The concept of LHM was first developed by a Russian physicist Victor Veselago in 1967 [4]. He predicted the existence of materials that could allow the propagation of electromagnetic waves with an electric field, magnetic field and phase vector that form a left-handed triad opposite to the commonly known right-handed triad of conventional materials (RHM). He also demonstrated that such materials are allowed by the Maxwell's equations. The most important electromagnetic properties discovered by Veselago are the reversal of Snell's Law, the reversal of the Doppler effect and the reversal of Vavilov-Cerenkov radiation. The cuspid of his work is not represented by the use of LHM in 1-dimensional structures but furthermore

M. Puentes Vargas, *Planar Metamaterial Based Microwave Sensor Arrays for Biomedical Analysis and Treatment*, Springer Theses,
DOI: 10.1007/978-3-319-06041-5_2,

in the realization of 2-dimensional and 3-dimensional isotropic and homogeneous media that can support backward waves and have a negative refractive index. Further analysis on his work can be found in [5] and [6].

Since Metamaterials are artificial materials and some of its properties are not found in nature, it is important that the Maxwell's equations are satisfied to obtain a physical meaningful result. The Maxwell's equations in its differential form are given as follows

$$\nabla \times \bar{E} = \frac{-\partial \bar{B}}{\partial t} - \bar{M} \tag{2.1}$$

$$\nabla \times \bar{H} = \frac{-\partial \bar{D}}{\partial t} + \bar{J} \tag{2.2}$$

$$\nabla \times \bar{D} = \rho \tag{2.3}$$

$$\nabla \times \bar{B} = 0 \tag{2.4}$$

where $\bar{E}(V/m)$ is the electric field density, $\bar{H}(A/m)$ is the magnetic field density, $\bar{D}(C/m^2)$ is the electric flux density, $\bar{B}(W/m^2)$ is the magnetic flux density, $\bar{M}(V/m^2)$ is the (fictitious) magnetic current density, $\bar{J}(A/m^2)$ is the electric current density and $\rho(C/m^3)$ is the electric charge density. If the medium is linear, non-dispersive, homogeneous and isotropic additional constitutive relations are given as follows

$$\bar{D} = \varepsilon_0 \bar{E} + \bar{P} = \varepsilon_0 (1 + \chi_e) \bar{E} = \varepsilon_0 \varepsilon_r \bar{E} = \varepsilon \bar{E} \tag{2.5}$$

$$\varepsilon_r = \varepsilon' - j\varepsilon'' \tag{2.6}$$

$$\bar{B} = \mu_0 (\bar{H} + \bar{M}) = \mu_0 (1 + \chi_m) \bar{H} = \mu_0 \mu_r \bar{H} = \mu \bar{H} \tag{2.7}$$

$$\mu_r = \mu' - j\mu'' \tag{2.8}$$

where $\bar{P}$ and $\bar{M}$ are the electric and magnetic polarization, χ_e and χ_m are the electric and magnetic susceptibility, ε_0 and μ_0 are the permittivity and permeability of free space and ε_r and μ_r are the permittivity and permeability of the material in question.

By considering a harmonic time-dependent plane wave traveling in the x-direction with electric field polarized in z-direction in a homogeneous, isotropic and linear medium, the Helmholtz wave equation looks as follows [7]

$$\frac{\partial^2}{\partial x^2} E_z(x) + \omega^2 \mu\varepsilon \, E_z(x) = 0 \tag{2.9}$$

$$\frac{\partial^2}{\partial y^2} H_y(x) + \omega^2 \mu\varepsilon \, H_y(x) = 0. \tag{2.10}$$

Assuming harmonic fields with time dependance $e^{j\omega t}$, the solution of $E_z(x)$ and $H_y(x)$ can be given as

$$E_z(x) = C_1 e^{-j\gamma x} + C_2 e^{j\gamma x} \tag{2.11}$$

$$H_y(x) = \frac{1}{\eta}(C_1 e^{-j\gamma x} + C_2 e^{j\gamma x}), \tag{2.12}$$

with

$$\gamma = \pm j\omega\sqrt{\mu\varepsilon} \tag{2.13}$$

$$\eta = \sqrt{\frac{\mu}{\varepsilon}}, \tag{2.14}$$

where η (Ω) is the wave impedance for the plane wave, defined as the ratio of the $\bar{E}$ and $\bar{H}$ fields and γ $(1/m)$ is the propagation constant. Since any physical scenario can be expressed as the superposition of plane waves of different frequencies according to the Fourier transform theorem, by considering a single plane wave the necessary information of the fundamental response of the medium can be obtained.

When considering a lossless medium and assuming that both μ and ε are positive, the propagation constant is purely imaginary and the plane wave will propagate outward from the source. When considering a more general case where μ and ε can take any value, including a simultaneous negative value, the propagation constant behaves in a more complicated fashion and both real and imaginary part of it has to be taken into account and it is given by

$$\gamma = \alpha + j\beta = \pm j\omega\sqrt{\mu\varepsilon}\sqrt{1 - j\frac{\sigma}{\omega\varepsilon}}, \tag{2.15}$$

where σ is the conductivity. Since the interest is on analyzing if LHMs are viable in terms of causality and therefore be used to construct periodic structures, the case of having both μ and ε negative will be addressed. By inserting this consideration into Eq. 2.15, one can notice that the attenuation constant α remains undisturbed as compared to having both variables positive, the interesting change occurs in the phase constant β. The β for RHM is positive and for LHM is negative. This results in a forward (away from the source) travel of the equiphase fronts for the RHM and a backward (towards the source) for the LHM. To satisfy causality in the LHM, it should be carefully analyzed what exactly propagates towards the source. The propagation of energy is not necessarily the same as the propagation of the equiphase fronts, for LHM they have opposite directions, this means, that energy will always propagate forward while the equiphase fronts move backwards. This can be seen in the phase velocity v_p and group velocity v_g as follows

$$v_p = \frac{\omega}{\beta} \tag{2.16}$$

$$v_g = \frac{\partial \omega}{\partial \beta}, \tag{2.17}$$

and the propagation characteristics for both RHM and LHM can be summarized by

$$\text{RHM}: \beta > 0, v_p > 0, v_g > 0 \tag{2.18}$$

$$\text{LHM}: \beta < 0, v_p < 0, v_g > 0. \tag{2.19}$$

Although the idea of a backward propagation seems troubling at first, it takes an interesting meaning once is realized that the phase velocity corresponds only to the propagation of a perturbation and not of energy itself. A negative group velocity would on the other hand be completely unpractical and would violate the law of causality. Since it has been proven that LHM can be fabricated [2] and be used to interact with electromagnetic waves, in the following sections will be addressed two different ways to physically use them and analyze them. The first one is as a transmission line structure and the second is an arrangement of resonant particles. Both approaches are useful for specific applications and are used throughout the work presented here in several different sensing scenarios. For this reason a comprehensible analysis of both approaches will be presented in this chapter.

2.2 Composite Right/Left Handed Transmission Line

Metamaterials are effective homogeneous materials that under certain assumptions can be modeled as a transmission line. They were first introduced in [1] and [3]. To consider them as a homogeneous material the key assumption is that a left-handed unit cell has an incremental length dz and the following restriction has to be applied

$$dz \ll \lambda_g = \frac{2\pi}{\beta}, \tag{2.20}$$

where λ_g represents the guided wavelength. In Fig. 2.1 is shown a section of an ideal transmission line. Inside the series impedance Z' and shunt inductance Y' different arrangement of lumped elements can be included. Two well known are the pure right-handed transmission line (PRH-TL) presented in Fig. 2.2a and the pure left-handed transmission line (PLH-TL) depicted in Fig. 2.2b. As mentioned before the LH-TL has some unique properties that can offer interesting features. Unfortunately in reality it is impossible to construct a PLH-TL since parasitic effects are unavoidable. The result is the composite right-left handed transmission line (CRLH-TL) shown in Fig. 2.2c.

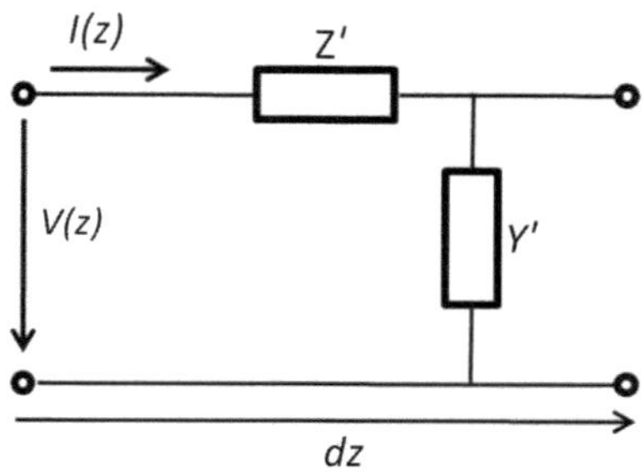

Fig. 2.1 Ideal homogeneous transmission line segment in the form of its incremental model

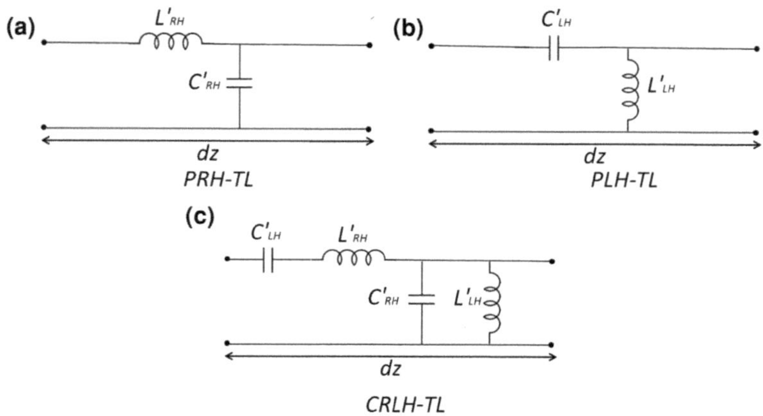

Fig. 2.2 Unit cell configurations: **a** PRH-TL, **b** PLH-TL and **c** CRLH-TL [5]

The propagation constant of any TL and the line impedance is given by [7]

$$\gamma = \alpha + j\beta = \sqrt{Z'Y'} \tag{2.21}$$

$$Z_L = R_L + jX_L = \sqrt{\frac{Z'}{Y'}} \tag{2.22}$$

where Z' and Y' are the per-unit length impedance and admittance respectively. For the CRLH-TL case they look as follows

$$Z' = j\left(\omega L'_{RH} - \frac{1}{\omega C'_{LH}}\right) \tag{2.23}$$

$$Y' = j\left(\omega C'_{RH} - \frac{1}{\omega L'_{LH}}\right). \tag{2.24}$$

By inserting Eqs. 2.23 and 2.24 into Eqs. 2.21 and 2.22 we obtain [8]

$$\gamma^{CRLH} = \pm j\omega\sqrt{\left(L'_{RH} - \frac{1}{\omega^2 C'_{LH}}\right)\left(C'_{RH} - \frac{1}{\omega^2 L'_{LH}}\right)} \tag{2.25}$$

$$Z_L^{CRLH} = \pm\sqrt{\frac{L'_{RH} - \frac{1}{\omega^2 C'_{LH}}}{C'_{RH} - \frac{1}{\omega^2 L'_{LH}}}}, \tag{2.26}$$

where Z_{LH} is the PLH impedance

$$Z_{LH} = \sqrt{\frac{L'_{LH}}{C'_{LH}}} \tag{2.27}$$

and Z_{RH} is the PRH impedance

$$Z_{RH} = \sqrt{\frac{L'_{RH}}{C'_{RH}}}. \tag{2.28}$$

At low frequencies the RH contribution will tend to zero and the line will behave as a PLH-TL, at high frequencies the LH contribution will tend to zero and the line will behave as a PRH-TL. At all other frequencies, the transmission behavior would be a combination of the RH and LH components and will result in a band gap. The limits of the band gap can be obtained with the correspondent series and parallel resonance frequencies f_s and f_p as follows

$$f_s = \frac{\omega_s}{2\pi} = \frac{1}{2\pi\sqrt{L'_{RH}C'_{LH}}} \tag{2.29}$$

$$f_p = \frac{\omega_p}{2\pi} = \frac{1}{2\pi\sqrt{L'_{LH}C'_{RH}}}. \tag{2.30}$$

2.2.1 Balanced and Unbalanced Conditions

Balanced Case

The CRLH-TL can be designed to have no band gap and therefore is said to be balanced when the following conditions are met

$$f_p \overset{!}{=} f_s \Rightarrow L'_{RH}C'_{LH} = L'_{LH}C'_{RH} \tag{2.31}$$

or equivalent

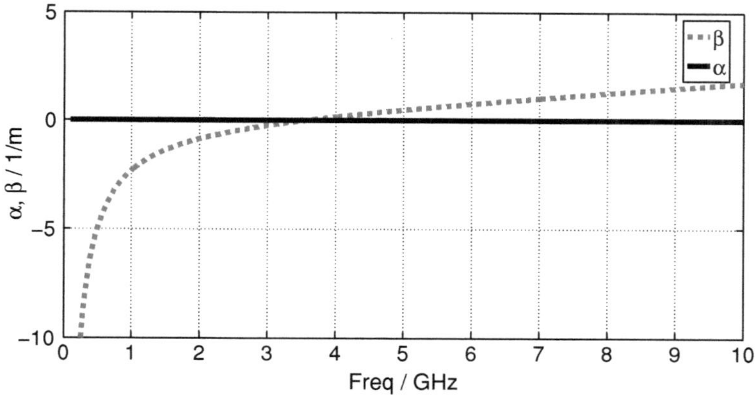

Fig. 2.3 Propagation constant of an exemplary ideal balanced CRLH line ($C_{LH} = 2\,\text{pF}$, $L_{RH} = 1\,\text{nH}$, $C_{RH} = 1\,\text{pF}$, $L_{LH} = 2\,\text{nH}$)

$$Z_c \stackrel{!}{=} Z_{LH} \stackrel{!}{=} Z_{RH}. \tag{2.32}$$

The propagation constant of a balanced CRLH-TL is plotted in Fig. 2.3. The unit cell is made of the following elements: $C_{LH} = 2\,\text{pF}$, $L_{RH} = 1\,\text{nH}$, $C_{RH} = 1\,\text{pF}$, $L_{LH} = 2\,\text{nH}$. Since the series and parallel resonance frequencies are equal, the pole and the zero that were present at f_s and f_p respectively, which have delimited the stop band, cancel each other. As a result, the stop band disappears and a continuous transmission is possible. The zero transition of the phase constant occurs at the transition frequency

$$\omega_0 = \omega_s = \omega_p = \frac{1}{\sqrt{L'_{RH} C'_{LH}}} = \frac{1}{\sqrt{L'_{LH} C'_{RH}}}. \tag{2.33}$$

In regard to the line impedance, it becomes purely real and constant with no dispersive behavior as depicted in Fig. 2.4, resulting in a good matching over a broad frequency spectrum. It can be represented by

$$Z_L = \sqrt{\frac{L'_{RH}}{C'_{RH}}} = \sqrt{\frac{L'_{LH}}{C'_{LH}}}. \tag{2.34}$$

Unbalanced Case

The CRLH-TL is said to be unbalanced when the following condition is met

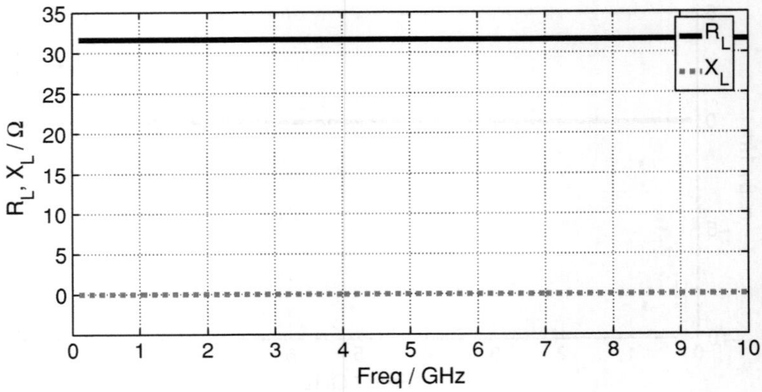

Fig. 2.4 Line impedance of an exemplary ideal balanced CRLH line ($C_{LH} = 2\,\text{pF}$, $L_{RH} = 1\,\text{nH}$, $C_{RH} = 1\,\text{pF}$, $L_{LH} = 2\,\text{nH}$)

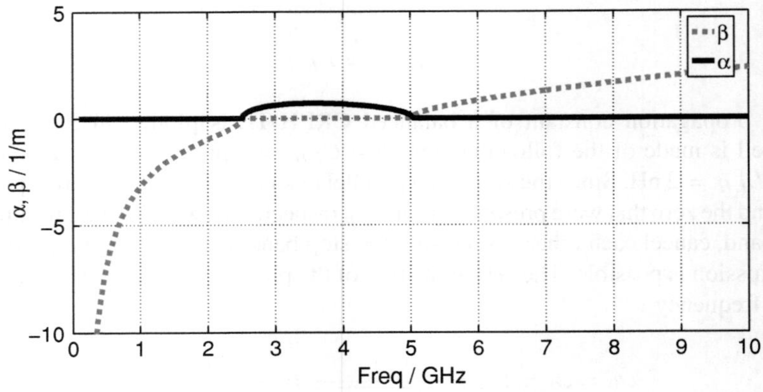

Fig. 2.5 Propagation constant of an exemplary ideal unbalanced CRLH line ($C_{LH} = 1\,\text{pF}$, $L_{RH} = 2\,\text{nH}$, $C_{RH} = 2\,\text{pF}$, $L_{LH} = 1\,\text{nH}$)

$$f_p \neq f_s. \tag{2.35}$$

The propagation constant of an unbalanced CRLH-TL is plotted in Fig. 2.5 and the line impedance can be observed in Fig. 2.6. The unit cell is made of the following elements: $C_{LH} = 1\,\text{pF}$, $L_{RH} = 2\,\text{nH}$, $C_{RH} = 2\,\text{pF}$, $L_{LH} = 1\,\text{nH}$. As a result a band gap is present between the frequencies f_s and f_p for the homogeneous line. In the band gap, the propagation constant takes real values and outside of the band gap imaginary values. Therefore Z' and Y' at these frequencies will have zero immittances introducing a zero and a pole in the characteristic impedance resulting in a zero group velocity. This means, that although a lossless case is assumed, the stop band or gap is present.

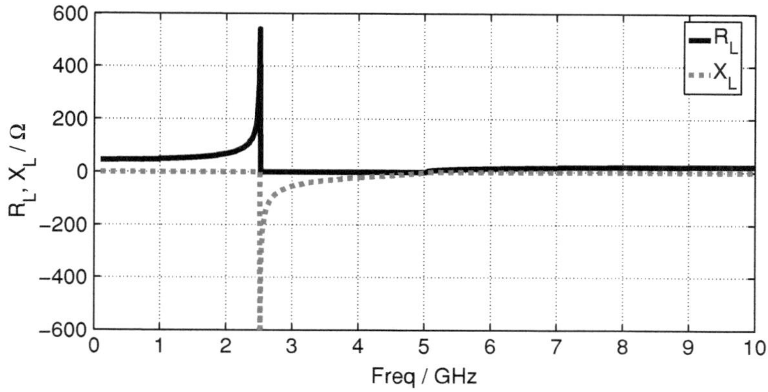

Fig. 2.6 Line impedance of an exemplary ideal unbalanced CRLH line ($C_{LH} = 1\,\text{pF}$, $L_{RH} = 2\,\text{nH}$, $C_{RH} = 2\,\text{pF}$, $L_{LH} = 1\,\text{nH}$)

The line impedance needs a more careful analysis. It has purely imaginary values in the band gap and purely real outside of it. At the frequencies that limit the band gap f_s and f_p, a pole and a zero can exist than can be interchanged by rotating the dimensionless unit cell values. As a result, two CRLH-TL can be constructed with identical resonance frequencies, identical propagation constant and the difference can only be seen in the impedance where it can take positive or negative values inside the band gap. This is important and can be used as an advantage for certain applications where both bands are not needed and a perfect matching can be obtained on the desired band by optimizing the design with the values of the unit cell elements.

2.2.2 Discrete Network Implementation

The ideal CRLH-TL previously described cannot be constructed, a real buildable CRLH-TL can be made by discretizing a homogeneous line in the form of a ladder network circuit by the transition of an infinitesimal unit cell of length dz to a finite unit cell length Δz. In the limit when $\Delta z \to 0$ the CRLH-TL will behave equivalent to the ideal case. To achieve this limit in practice, the size of the unit cell should be smaller than the guided wavelength $\Delta z < \lambda_g/4$ to obtain an electrical length smaller than $\pi/2$. The discretized values of the inductors and capacitors look as follow

$$C_{LH} = C'_{LH}/\Delta z \tag{2.36}$$
$$L_{RH} = L'_{RH} \cdot \Delta z \tag{2.37}$$
$$C_{RH} = C'_{RH} \cdot \Delta z \tag{2.38}$$
$$L_{LH} = L'_{LH}/\Delta z. \tag{2.39}$$

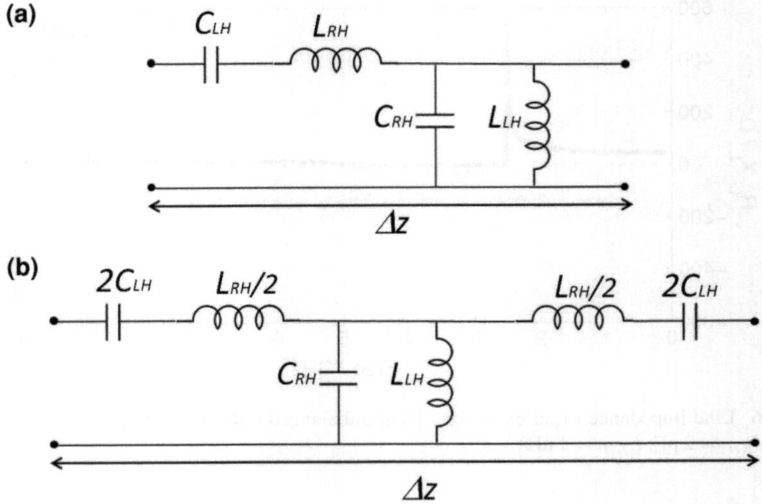

Fig. 2.7 Discrete CRLH unit cell: **a** Asymmetric and **b** Symmetric

Since the transmission line is not infinitely long, it has to be taken into account how the line is terminated. In Fig. 2.7 is shown the case when the CRLH unit cell is terminated in an asymmetric or symmetric way.

For simplicity, the symmetric unit cell will be analyzed and similar analysis can also be conducted for the asymmetric unit cell. The propagation characteristics, namely propagation constant and Bloch impedance, looks as follows [8]

$$\begin{aligned}\gamma &= \pm\frac{1}{\Delta z}\cosh^{-1}\left(1+\frac{ZY}{2}\right)\\ &= \pm\frac{1}{\Delta z}\cosh^{-1}\left(1-\frac{\omega^2}{2}\left(L_{RH}-\frac{1}{\omega^2 C_{LH}}\right)\left(C_{RH}-\frac{1}{\omega^2 L_{LH}}\right)\right)\end{aligned} \tag{2.40}$$

$$\begin{aligned}Z_B &= \pm Z\sqrt{\frac{1}{ZY}+\frac{1}{4}}\\ &= \pm j\omega^2\left(L_{RH}-\frac{1}{\omega^2 C_{LH}}\right)\sqrt{\frac{1}{4}-\frac{1}{\omega^2\left(L_{RH}-\frac{1}{\omega^2 C_{LH}}\right)\left(C_{RH}-\frac{1}{\omega^2 L_{LH}}\right)}}\end{aligned} \tag{2.41}$$

It should be noted that for both symmetric and asymmetric unit cells the propagation constant is identical and the difference can be seen only in the Bloch impedance. The Bloch impedance represents the characteristic impedance of the discrete implementation of the LC network and takes into account the case where $\Delta z > 0$. In Figs. 2.8

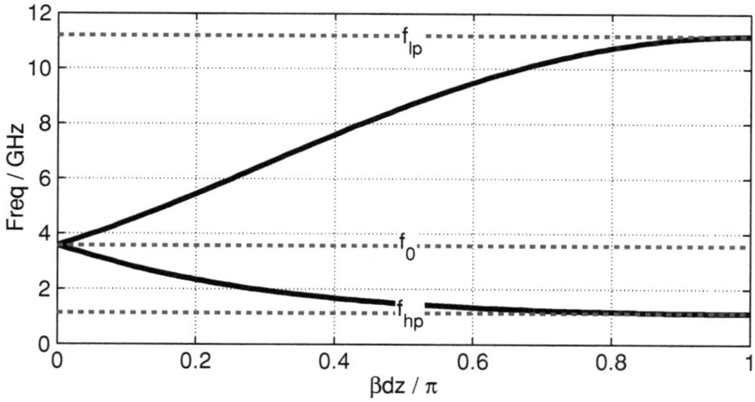

Fig. 2.8 Dispersion diagram for the discrete balanced symmetric CRLH-TL

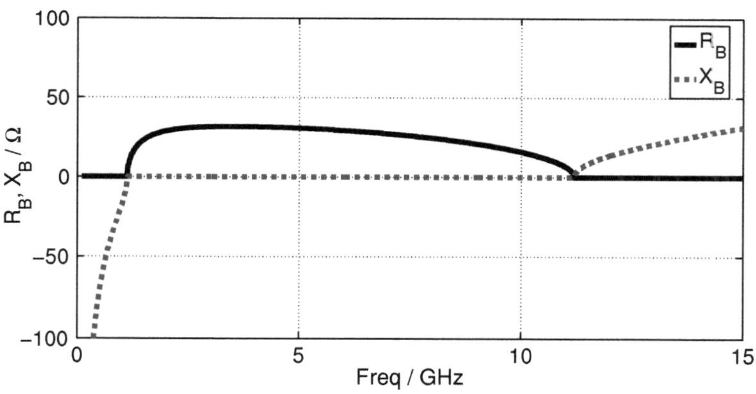

Fig. 2.9 Bloch impedance for the discrete balanced symmetric CRLH-TL

and 2.9 is shown the dispersion diagram and Bloch impedance for the balanced case with lumped elements $C_{LH} = 2\,\text{pF}$, $L_{RH} = 1\,\text{nH}$, $C_{RH} = 1\,\text{pF}$, $L_{LH} = 2\,\text{nH}$. The transition frequency can be calculated with

$$\omega_0 = 2\pi f_0 = \frac{1}{\sqrt[4]{L_{RH} C_{LH} L_{LH} C_{RH}}}. \tag{2.42}$$

The dispersion diagram of a discretized balanced CRLH-TL presented in Fig. 2.8 shows a big difference to the ideal CRLH-TL that can be immediately recognized. The LH and RH bands do not extend to infinity but experience a cutoff frequency f_{hp} at lower frequencies and f_{lp} at higher frequencies. Outside of these boundaries the discrete line will lose the properties of the homogeneous line since the discretization fails. These cutoff frequencies can be obtained by inspection of the input impedance and are given as follows [8]

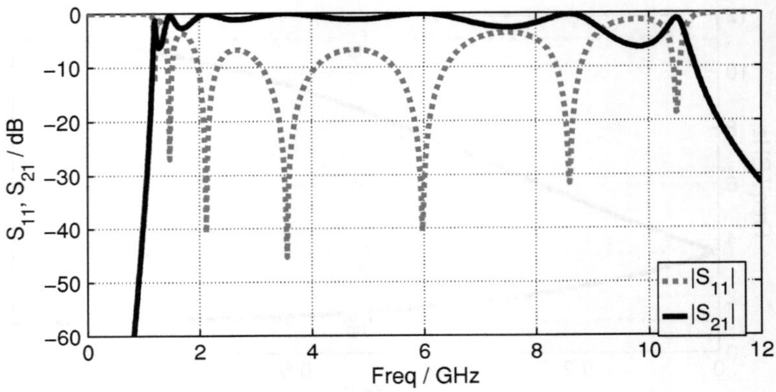

Fig. 2.10 S-parameters for a discrete balanced symmetric CRLH-TL with four identical unit cells ($C_{LH} = 2\,\text{pF}$, $L_{RH} = 1\,\text{nH}$, $C_{RH} = 1\,\text{pF}$, $L_{LH} = 2\,\text{nH}$)

$$f_{hp} = \frac{1}{2\pi}\sqrt{p/2 - \sqrt{p^2/4 - q}} \tag{2.43}$$

$$f_{lp} = \frac{1}{2\pi}\sqrt{p/2 + \sqrt{p^2/4 - q}} \tag{2.44}$$

with

$$p = \frac{1}{C_{RH}L_{LH}} + \frac{1}{L_{RH}C_{LH}} + \frac{4}{L_{RH}C_{RH}} \tag{2.45}$$

$$q = \frac{1}{L_{RH}C_{LH}L_{LH}C_{RH}}. \tag{2.46}$$

The cutoff frequencies effect can also be observed in the Bloch impedance where the impedance is no longer real and constant as with the homogeneous balanced case but becomes purely imaginary outside of the boundaries of the cutoff frequencies.

Finally, in Fig. 2.10 the input reflection S_{11} and forward transmission S_{21} of a discrete symmetric and balanced CRLH-TL are shown. It has four identical unit cells. From the figure can be seen once more the cutoff frequencies f_{lp} and f_{hp} that delimit the transmission band. The influence of the Bloch impedance can also be appreciated in the input reflection S_{11}, the dips represent the points where the terminating port is matched to the input port. Theoretically there should be four peaks on the left handed band, four on the right handed band and one on the transition frequency since we have a balanced structure. In the figure only three peaks can be observed on each band. The missing dips correspond to the cutoff frequencies, at these values the Bloch impedance becomes zero and therefore the input impedance is not equal to the port impedance but takes a zero value. If the structure was unbalanced the dip at the transition frequency would disappear and a band gap would be there instead.

The metamaterial transmission lines will be used in different sensing applications that will be presented in the following chapters. The reason for using them is the possibility of constructing compact sensors that can operate at lower frequencies than with conventional lines. On the other hand, their periodicity can be combined with extraction methods to create a simultaneous monitoring of several samples of material under test (MUT). Finally, even though any capacitance can be used as the sensing element, the most straightforward is the C_{LH} due to its location in the serial branch, especially for planar applications which is the main focus of this work. Additionally, under certain design conditions the C_{LH} can perform with a higher sensitivity than the C_{RH} as will be seen in a comprehensive analysis between them, specifically for sensing applications, in Sect. 4.2.

2.3 Resonant Metamaterial Structures

Additionally to the metamaterial structures using the transmission line approach, there is another type of metamaterial structures based on sub-wavelength resonators, mainly using split ring resonator (SRR). The original topology was first proposed by Pendry [2]. Several different approaches of resonant particles based on SRR have been developed and applied for a number of microwave applications. Their advantages lie on the fact that they are very compact structures with high flexibility in the design process making them suitable for miniaturization purposes. They can be coupled in many different ways by combining them with coplanar, stripline and microstrip designs adding the planarity feature. Depending on the type of resonator and coupling used, different propagation characteristics can be tailored to the desired requirements. It should be emphasized that this approach is named resonant approach because resonators are used as loading elements. The transmission characteristics are qualitatively similar to the CRHL-TL approach presented in the previous section, that means, the resonant approach also exhibits CRLH behavior and can be designed to be balanced and have no stop band. The difference between the two approaches can be visualized on their equivalent circuit models.

2.3.1 Split Ring Resonator

A simple edge coupled SRR is formed by two concentric open metallic rings as depicted in Fig. 2.11. They can have a circular or a square shape. The SRR can be excited by an axial time-varying external magnetic field, which induces current between them. For example, by placing them in the proximity of a microstrip line, the magnetic field present in the line will couple to the SRR. The current loop is closed by the distributed capacitance that appears between the outer and inner ring and the gaps of both of them. They can be modeled as a resonant tank as shown in Fig. 2.11c [9].

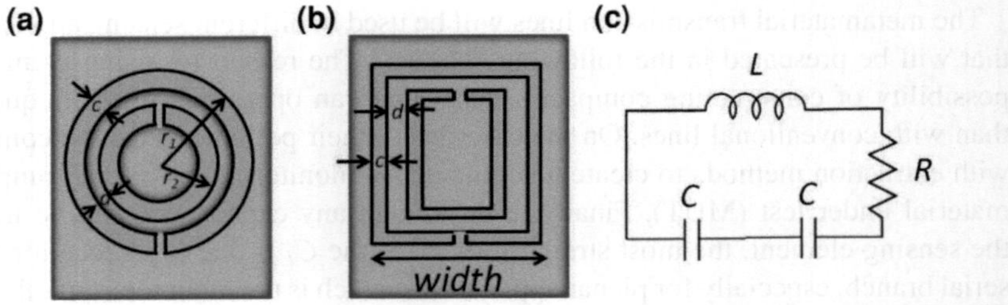

Fig. 2.11 SRR and its most important dimensions: **a** circular design, **b** square design and **c** equivalent circuit

The equivalent circuit has a self-inductance L and a capacitance C associated with each half of the SRR represented as follows

$$C = \pi r C_{pul}, \tag{2.47}$$

where r is the mean radius of the SRR ($r = r_1 - c - d/2$) and C_{pul} is the per-unit length capacitance along the slot between the rings. The total capacitance is the sum of both SRR halves, that is $C/2$. Neglecting losses the total current I is represented as [9]

$$\left(\frac{2}{j\omega C} + j\omega L\right) I = \Sigma \tag{2.48}$$

where Σ is the external excitation. The Eq. 2.48 is valid as long as the perimeter of the ring is smaller than half-wavelength and the capacitance at the gaps neglected so the currents at this place are vanished, therefore, the currents on each ring are assumed to be linear. The voltage across the slots is also constant and the resonant frequency ω_0 can be calculated by solving Eq. 2.48 for $\Sigma = 0$ as follows [9]

$$\omega_0^2 = \frac{2}{LC} = \frac{2}{\pi r C_{pul} L}. \tag{2.49}$$

The main goal of this work is to use an array of SRRs as will be discussed in Sect. 6.2. The SRRs have to be decoupled and detuned from each other in order to have different resonant frequencies and behave independently from each other. From Eq. 2.49 can be deducted that this can be done by changing the size of the SRRs. By changing for example the radius of the SRR the frequency shifts. To simplify the process, it was decided to change only one parameter and the selected one was the radius or width of the SRR particle depending if the are circular or square. In Fig. 2.12 it is shown how the resonant frequency of a particular square ring is affected by changing the width from 0.3 to 4 mm. This proofs that an array of independent SRRs can be done by changing their geometry. This will be very useful in the intended application of this work.

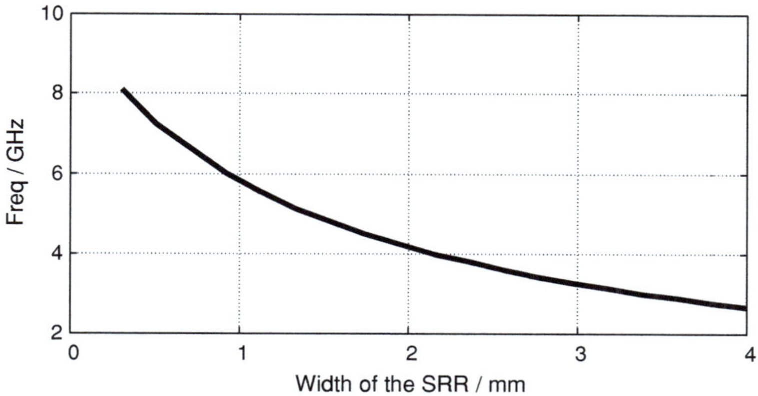

Fig. 2.12 Dependance of SRR resonance frequency f_0 on the width of the SRR

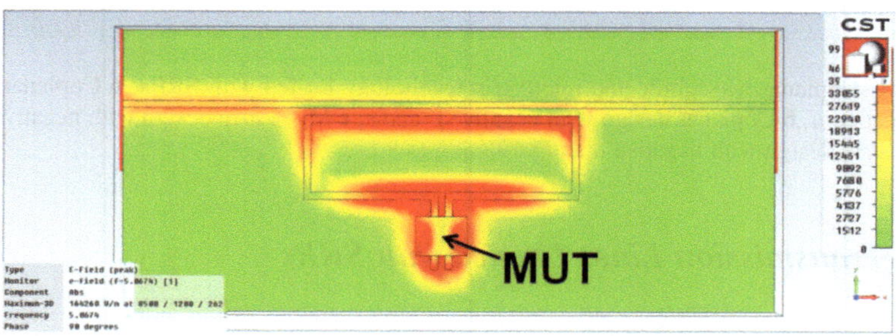

Fig. 2.13 Field distribution on one microstrip excited SRR loaded with a material under test (MUT). The field is clearly stronger around the gap of the ring where the MUT is located and around the coupling to the feeding microstrip line

As it was mentioned in the introduction, another goal of the array of SRRs is to perform dielectric analysis and thermal ablation treatment of organic tissue. For this purpose it is important to analyze the field distribution in order to see if the energy can be focused around the outer gap of the SRR. In Fig. 2.13 an example of a microstrip excited SRR that is in contact with a MUT of $\varepsilon_r = 10$ is depicted. The field distribution is shown and it concentrates around the SRR particle, especially around the MUT where the field has a strength 10 times higher than in the rest of the structure. This ensures an appropriate interaction that will provide the necessary information for making an adequate extraction of the dielectric image of the MUT. Furthermore it ensures a good coupling of the energy into the tissue to be able to perform thermal ablation treatment on the material as will be seen in Chap. 6.

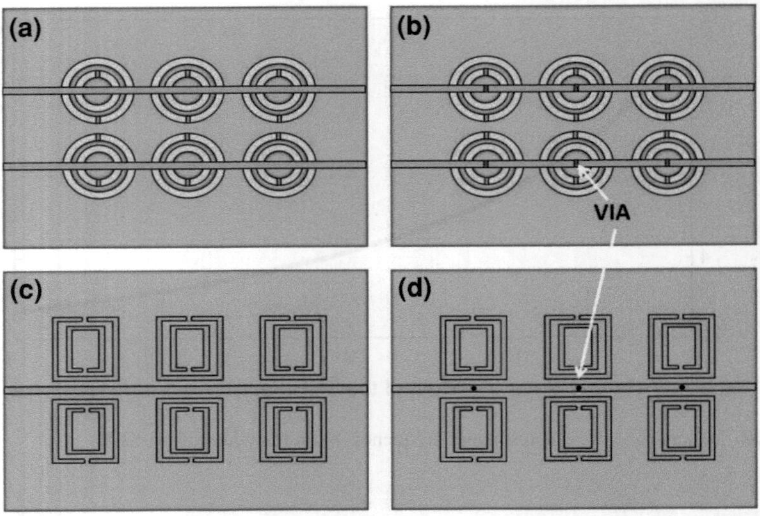

Fig. 2.14 Coplanar waveguide and microstrip technology loaded with SRRs: **a** Coplanar design with negative μ, **b** Coplanar design with negative μ and ε, **c** Microstrip design with negative μ and **d** Microstrip design with negative μ and ε [9]

2.3.2 Transmission Lines Loaded with SRRs

When a periodic array of SRRs is considered, it is possible to obtain either signal propagation or inhibition in a narrow band near the resonance frequency of the SRRs. They behave as lumped resonators and can be designed to have a size much smaller than the wavelength. Due to the resonance present in the SRR medium, it will have extreme permeability both positive and negative around the resonance frequency and therefore no propagation occurs at this frequency. Additionally by including metallic posts or gaps, a simultaneous negative permittivity can also be achieved, i.e. a LHM can be fabricated in a narrow band above the SRR resonance frequency [9].

In Fig. 2.14 four different geometries to excite SRRs are depicted. The first scenario is shown in Fig. 2.14a where a coplanar waveguide is used to excite the SRRs. When a coplanar waveguide is used for this purpose, the SRRs can be etched on the upper or lower side of the substrate for a uniplanar or bi-metal implementation, respectively. The bi-metal choice is preferred since it requires less space to accommodate the SRR and it also reduces the influence on the impedance matching with the line allowing for a better coupling between them. Therefore the bi-metal implementation has a higher coupling and low insertion losses in the desired frequency band [9]. In Fig. 2.14a it is also seen that the SRRs have a slightly different color to represent that they are etched on the back side of the substrate. The resonant frequency can be calculated with Eq. 2.49 and for this specific example with port impedance of 50 Ω a stopband occurs above and below f_0 as seen in Fig. 2.15a. The reason for the

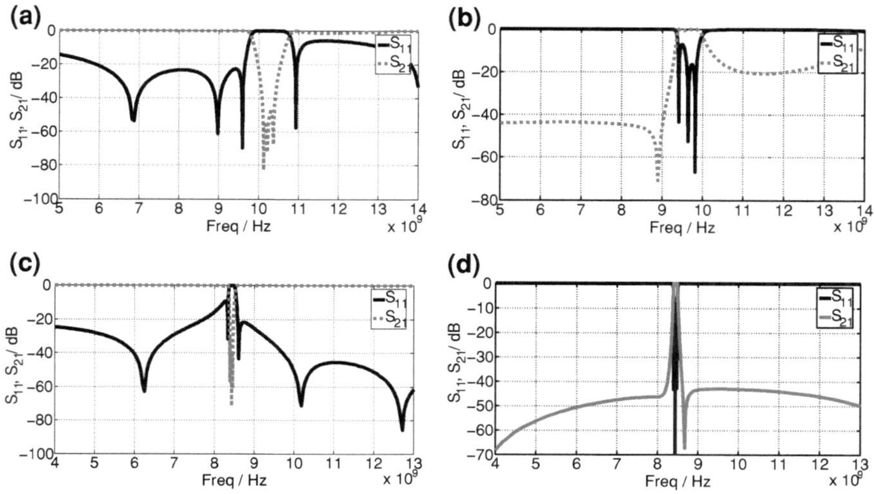

Fig. 2.15 S-parameters of the coplanar waveguide (conductor width = 2.5 mm, gap width = 0.35 mm, r_1 = 0.8 mm, r_2 = 0.4 mm and c = 0.2 mm) and microstrip line (conductor width = 0.2 mm, distance between conductor and rings = 0.2 mm, distance between adjacent rings 1 mm, width of each ring = 2 mm, c = 0.2 mm and d = 0.2 mm) loaded with SRRs: **a** Coplanar design with negative μ, **b** Coplanar design with negative μ and ε, **c** Microstrip design with negative μ and **d** Microstrip design with negative μ and ε (Substrate ε_r = 10.6 and thickness = 254 μm)

stopband is that above f_0 a negative magnetic permeability is present and below a positive permeability in a narrow band also exists resulting in a strong mismatch at the input port of the line. Outside of this frequency range the signal can propagate without any problem, it simply is not affected by the SRRs.

The second possibility corresponds to the first LH-TL implemented using SRR and was reported in [10]. It is a coplanar waveguide with backside etched SRRs and a signal to ground metallic connections (vias) periodically located above the SRRs as shown in Fig. 2.14b. It has simultaneously negative permeability (produced by the SRRs) and permittivity (produced by the vias) in a certain band due to the strips that behave as shunt inductors up to a certain frequency called the plasma or cutoff frequency f_c. As long as f_c is above f_0 a narrow band region will exist where an LH-TL operates above f_0. In this band, signal propagation exists but with backward waves, that is, with negative phase velocity. The propagation is exactly opposite to the data presented in Fig. 2.15a, where the stopband was located, a pass band now takes its place only slightly shifted since it starts exactly above f_0 as seen in Fig. 2.15b. The signal does not propagate outside this band.

The third scenario is done with a microstrip technology design as shown in Fig. 2.14c. The only choice in this particulary case is to etch the SRRs close to the microstrip line to obtain a high magnetic coupling supported by the square shaped SRRs, since the coupling area is larger compared to the circular rings. The propagation is exactly the same as described for the coplanar design of Fig. 2.14a. A

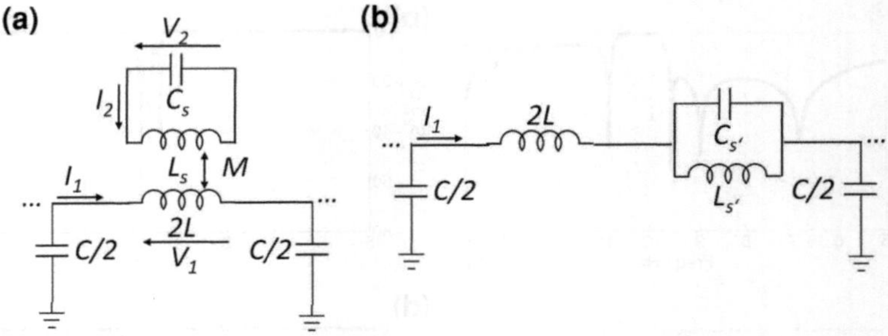

Fig. 2.16 **a** Equivalent circuit for a microstrip loaded with SRRs with negative permeability and **b** circuit model after transformation of the series branch [9]

stopband occurs near f_0 and a good transmission exist outside this band as seen in Fig. 2.15c. The last design is shown in Fig. 2.14d and it includes the microstrip line with square SRR and via holes are included to emulate again shunt inductors and obtain an LH-TL. The behavior is exactly the same as from the structure of Fig. 2.14b. The S-parameters for this case are presented in Fig. 2.15d. Through this work the design of Fig. 2.14c was mainly used so an additional analysis of this scenario will be given.

2.3.3 Microstrip Line Coupled with SRRs as Sensor Elements

The main target of transmission lines based on SRRs in this work is for biomedical applications where the monitoring of dielectric properties of cells or organic tissues is necessary. Furthermore they are used also as a treatment option for cancer tissue when the SRRs are able to heat the tissue and perform thermal ablation. It is not necessary in this case to have them behave as LHM but only as resonators. For this reason the design presented in Fig. 2.14c was selected but it should be emphasized that any of the designs presented in Fig. 2.14 are suitable for this application with its correspondent advantages and disadvantages. The equivalent circuit of the mentioned structure is given in Fig. 2.16a. The model includes a resonant tank to represent the SRR, the host transmission line and the magnetic coupling M between them. Although SRRs can be excited by both magnetic and electric fields, the magnetic coupling is the predominant mechanism and will be considered in the analysis. Since the design under consideration includes microstrip technology, the magnetic coupling cannot be easily calculated by a closed formula and is considered a fitting parameter [9].

The equations for the series and parallel impedance of the Fig. 2.16a are [11, 12]:

$$Z_{SRR}(\omega) = \frac{j\omega(-L + LC_sL_s\omega^2 - 2M^2C_s\omega^2)}{C_sL_s\omega^2 - 1} \tag{2.50}$$

$$Y_{SRR}(\omega) = \frac{2}{j\omega C}, \tag{2.51}$$

and for the transformed model presented in Fig. 2.16b, the equations look as follows [11, 12]:

$$\tilde{Z}_{SRR}(\omega) = \frac{j\omega(L' - L'\omega^2C_s'L_s' + L_s')}{1 - \omega^2C_s'L_s'} \tag{2.52}$$

$$\tilde{Y}_{SRR}(\omega) = \frac{2}{j\omega C} \tag{2.53}$$

with the following conditions

$$\tilde{L_s} = 2M^2C_s\omega_0^2 \tag{2.54}$$

$$\tilde{C_s} = \frac{L_s}{2M^2\omega_0^2} \tag{2.55}$$

$$\tilde{L} = L - \tilde{L_s}. \tag{2.56}$$

The Bloch impedance for both the original model and the transformed model looks as follows:

$$Z_B = 2\sqrt{\frac{-L + LC_sL_s\omega^2 - 2M^2C_s\omega^2}{C(-4 + 4C_sL_s\omega^2 + \omega^2CL - \omega^4CLC_sL_s + 2\omega^4CM^2C_s)}} \tag{2.57}$$

$$\tilde{Z_B} = 2\sqrt{\frac{-\tilde{L} + \tilde{L}\tilde{C_s}\tilde{L_s}\omega^2 - \tilde{L_s}}{C(-4 + 4\tilde{C_s}\tilde{L_s}\omega^2 + \omega^2C\tilde{L} - \omega^4C\tilde{L}\tilde{C_s}\tilde{L_s} + \omega^2C\tilde{L_s})}}. \tag{2.58}$$

The important frequencies to calculate in order to see if propagation takes place or not are the frequencies where $Z_s = \infty$ (ω_s) and $Z_s = 0$ (ω_0). For the model in Fig. 2.16a are:

$$Z_{SRR} = \infty \tag{2.59}$$

$$L_sC_s\omega^2 - 1 = 0 \tag{2.60}$$

$$\omega_s = \sqrt{\frac{1}{L_sC_s}} \tag{2.61}$$

$$Z_{SRR} = 0 \tag{2.62}$$

$$-L + LC_sL_s\omega^2 - 2m^2C_s\omega^2 = 0 \tag{2.63}$$

$$\omega_0 = \sqrt{\frac{L}{C_s(L_s L + 2M^2)}}. \tag{2.64}$$

And for the model in Fig. 2.16b are:

$$Z_{\tilde{SRR}} = \infty \tag{2.65}$$

$$1 - \tilde{L_s}\tilde{C_s}\omega^2 = 0 \tag{2.66}$$

$$\tilde{\omega_s} = \sqrt{\frac{1}{\tilde{L_s}\tilde{C_s}}} \tag{2.67}$$

$$Z_{\tilde{SRR}} = 0 \tag{2.68}$$

$$\tilde{L} - \tilde{L}\tilde{C_s}\tilde{L_s}\omega^2 + \tilde{L_s}\omega^2 = 0 \tag{2.69}$$

$$\tilde{\omega_0} = \sqrt{\frac{\tilde{L}}{\tilde{L_s}(\tilde{L}\tilde{C_s} - 1)}}. \tag{2.70}$$

Since this is the design that is mainly used in the applications presented in this work and for continuity to the way the analysis was made in the transmission line approach, the propagation characteristics of this model as well as the impedance behavior will be analyzed. The dispersion constant is given by [9]

$$\cos(\beta l) = 1 + \frac{Z_{SRR}(\omega)}{Y_{SRR}(\omega)} \tag{2.71}$$

where $Z_{SRR}(\omega)$ and $Y_{SRR}(\omega)$ are the series impedance and shunt admitance of the circuit model. By applying Eq. 2.71 to the specific equivalent circuit presented in Fig. 2.16, it transforms to [9]:

$$\cos(\beta l) = 1 - \frac{LC\omega^2}{2} + \frac{C/\tilde{C_s}}{4\left(1 - \frac{\omega_0^2}{\omega^2}\right)}. \tag{2.72}$$

From Eq. 2.72, it can be seen that the propagation characteristic changes if the capacitance of the SRR C_s' is affected by loading a material on the SRR. This feature will be thoroughly exploited in the biomedical applications presented in this work. To represent graphically the previous equations, the equivalent circuit of a single SRR coupled to a microstrip line was simulated. The values for the model are $C_s = 1\,\text{pF}$, $L_s = 1\,\text{nH}$, $C = 1\,\text{pF}$, $L = 1\,\text{nH}$ and the unit length taken into account for the extraction of the propagation constant is 5 mm. The coupling between the SRR and the host line was changed from $M = 0$ to represent no coupling to $M = 0.2$ for

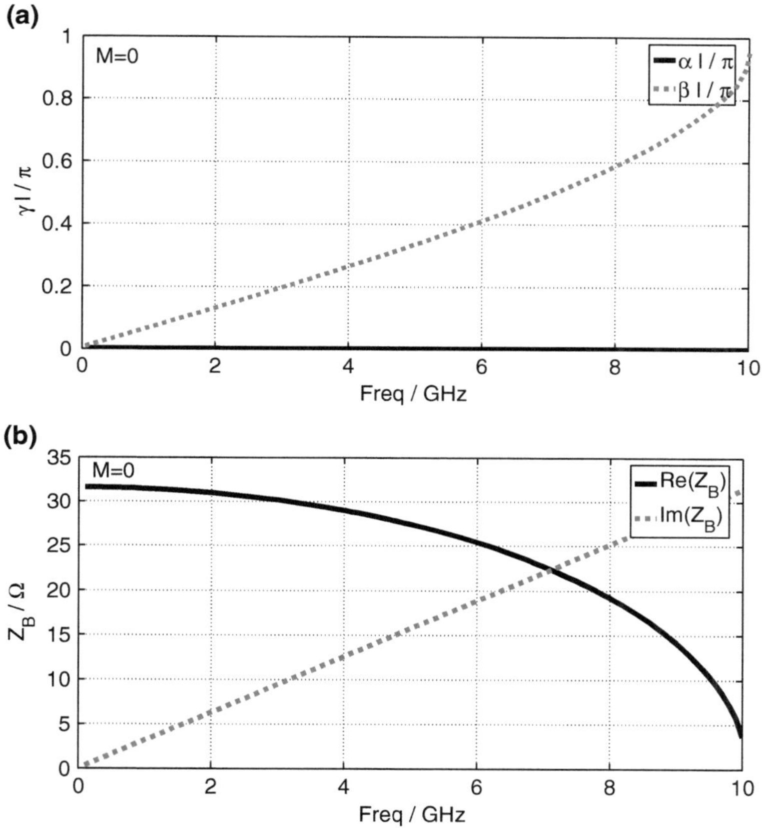

Fig. 2.17 **a** Propagation constant and **b** Bloch impedance of a microstrip line loaded with an SRR ($C_s = 1\,\text{pF}$, $L_s = 1\,\text{nH}$, $C = 1\,\text{pF}$, $L = 1\,\text{nH}$). The coupling between the line and the SRR is $M = 0$

medium coupling and finally to $M = 0.7$ for increased coupling. The results for each of the mentioned cases can be seen in Figs. 2.17, 2.18 and 2.19. Additionally, in these figures the correspondent Bloch impedances are also depicted.

The first scenario with coupling $M = 0$ shows how the host line simple does not see the SRR at all. There is no resonance present and a constant transmission in the frequency band occurs before the discretization starts to fail as for the transmission line approach. The Z_B also shows a semi constant behavior. In the second scenario where the coupling is slightly increased to $M = 0.2$, the host line starts to be influenced by the SRR and a resonance occurs around 5 GHz. In this case the ω_s and ω_0 are almost the same and the expected stop band is very small. This will result in a very sharp peak in the scattering parameters. As the coupling increases to $M = 0.7$, ω_s and ω_0 separate further from each other creating a bigger stop band. This would insert more energy in the structure but the peaks in the scattering parameters would be

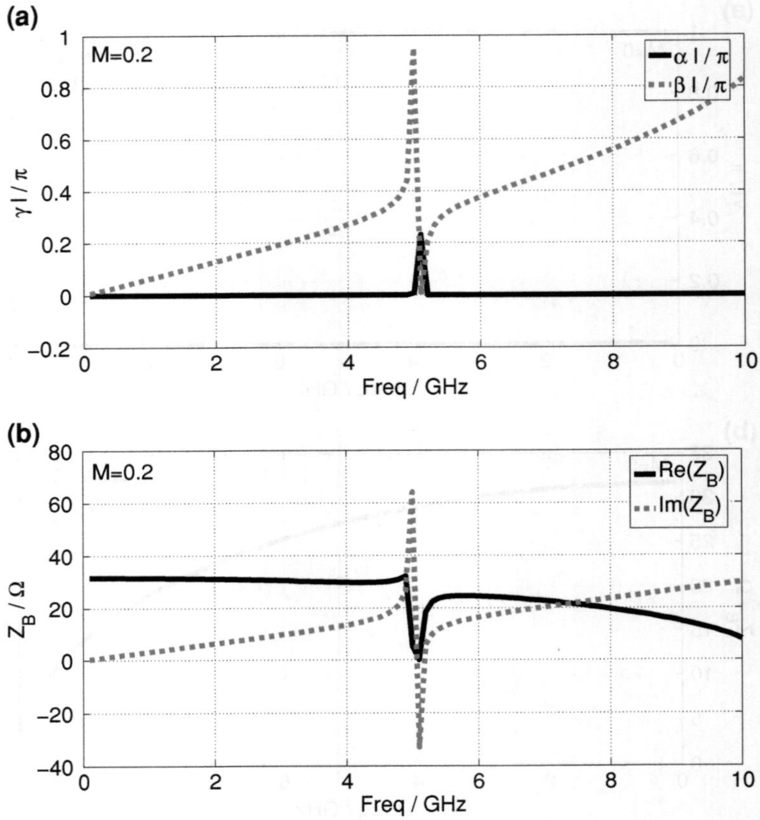

Fig. 2.18 **a** Propagation constant and **b** Bloch impedance of a microstrip line loaded with an SRR ($C_s = 1\,\text{pF}$, $L_s = 1\,\text{nH}$, $C = 1\,\text{pF}$, $L = 1\,\text{nH}$). The coupling between the line and the SRR is $M = 0.2$

wider. For the application presented in this work it is necessary to find a compromise in the coupling since is important to have sharp peaks for using the structure as a sensor but it is also necessary to insert a lot of energy to maximize the efficiency when used for thermal heating of the loading material.

To observe how a small array would behave, the simulated propagation constant and the normalized characteristic impedance for a prototype with two SRRs is shown in Fig. 2.20. A stop band for each SRR is produced in the transmission spectrum due to magnetic resonance for the shown configuration at 0.425 GHz and 1.04 GHz. When the gaps are loaded with a dielectric perturber which increases the overall capacitance, or equivalently the permittivity over the gap, a decrease in the resonance frequency of the loaded SRR will occurred.

Finally, the scattering parameters of an array of four SRRs with slightly different widths to decoupled them is shown in Fig. 2.21. As expected from what has been

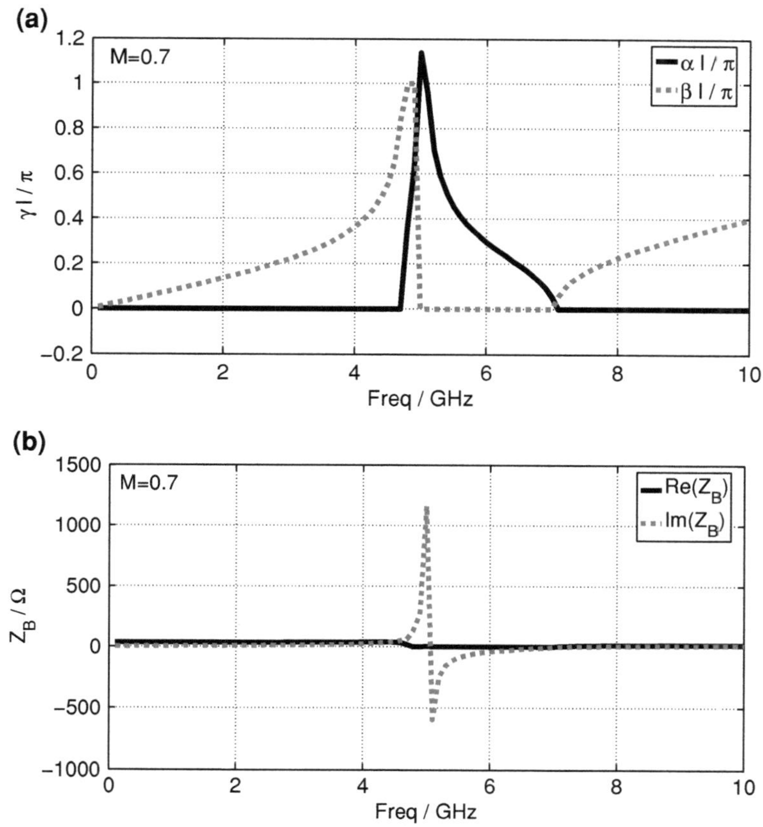

Fig. 2.19 **a** Propagation constant and **b** Bloch impedance of a microstrip line loaded with an SRR ($C_s = 1\,\text{pF}$, $L_s = 1\,\text{nH}$, $C = 1\,\text{pF}$, $L = 1\,\text{nH}$). The coupling between the line and the SRR is $M = 0.7$

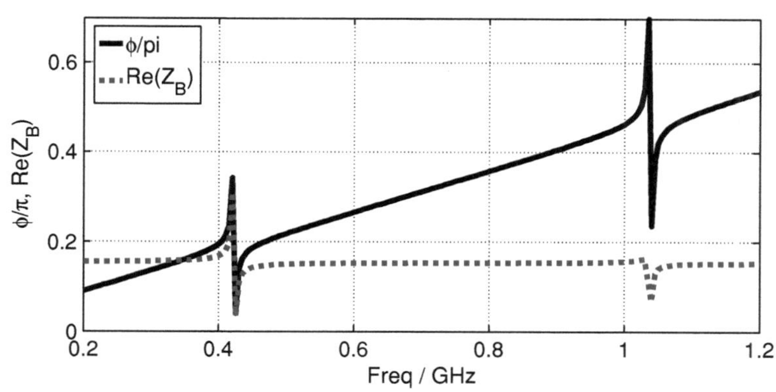

Fig. 2.20 Simulated propagation constant and normalized characteristic impedance of the prototype with two SRRs. Each SRR produces a separate stop band

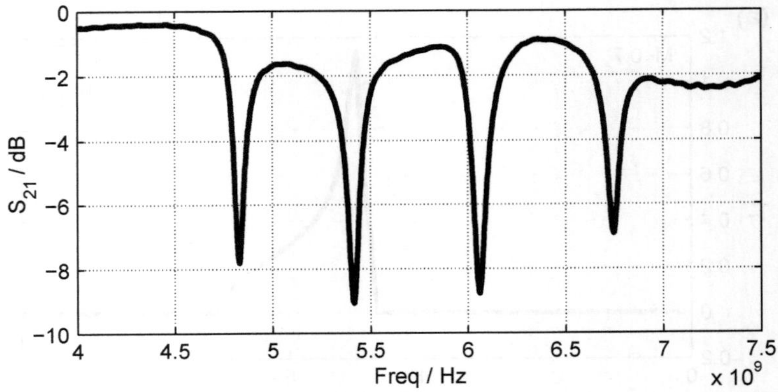

Fig. 2.21 S-parameters of an array of four SRRs. Four independent resonance frequencies can be observed

discussed in this chapter, four resonant peaks can be observed. All SRRs are decoupled and detuned from each other, they are able to interact with a loading material in an independent way and furthermore under a certain scenario they are able to focus the energy in their gaps. The complete discussion has been done without considering any losses coming from the structure itself since they are not necessary for the explanation of the operation principle. They are not an issue in the intended application, the only losses that cause difficulties are coming from the organic material itself as will be discussed in Chaps. 5 and 6.

In summary, metamaterial structures exhibits interesting properties that can be used for sensing applications. Their flexibility in terms of operation frequency, size and geometry can be exploited to create novel, cost-efficient sensor devices. In the following chapters will be studied a novel sensor concept based on this materials and it will be applied in different application scenarios to prove its wide functionality and adaptability.

References

1. C. Caloz, T. Itoh, Application of the transmission line theory of left-handed (LH) materials to the realization of a microstrip LH line, in *Proceedings of the IEEE Antennas and Propagation Society International Symposium*, June 2002. pp. 412–415
2. J. Pendry, A. Holden, D. Robbins, W. Stewart, Magnetism from conductors and enhanced nonlinear phenomena. IEEE Trans. Microw. Theory Tech. **47**(11), 2075–2084 (1999)
3. G. Eleftheriades, A. Iyer, P. Kremer, Planar negative refractive index media using periodically L-C loaded transmission lines. IEEE Trans. Microw. Theory Tech. **50**(12), 2702–2712 (2002)
4. V. Veselago, The electrodynamics of substances with simultaneously negative values of ε and μ. Sov. Phys. Usp. **10**(4), 509–514 (1968)
5. C. Caloz, T. Itoh, *Electromagnetic Metamaterials* (Wiley, Hoboken, 2006)
6. G. Eleftheriades, K. Balmain, *Negative-Refraction Metamaterials* (Wiley, Hoboken, 2005)

7. D. Pozar, *Microwave Engineering* (Wiley, Hoboken, 2005)
8. C. Damm, *Artificial Transmission Line Structures for Tunable Microwave Components and Microwave Sensors* (Shaker Verlag, Aachen, 2011)
9. R. Marques, F. Martin, M. Sorolla, *Metamaterial with Negative Parameters* (Wiley, Hoboken, 2008)
10. F. Martín, F. Falcone, J. Bonache, R. Marqués, M. Sorolla, Split ring resonator based left handed coplanar waveguide. Appl. Phys. Lett. **83**, 4652–4654 (2003)
11. F. Aznar Ballesta, Caracterización de Nuevos Resonadores Metamaterial, Líneas de Transmisión Artificiales y Aplicación en el Diseño de Circuitos de Comunicaciones, Ph.D. dissertation, Universitat Autònoma de Barcelona, Departament d'Enginyeria Electrònica, July 2009
12. F. Aznar Ballesta, Revisión del modelo circuital de una línea de transmisión metamaterial basada en split ring resonators (srrs), Master's thesis, Universitat Autònoma de Barcelona, Departament d'Enginyeria Electrònica, 2009

Chapter 3
Extraction Techniques

All the sensors that will be presented in this work have the purpose of extracting the dielectric properties of a MUT in order to obtain information about its constituents, position and recreate a dielectric image. The advantage of the methods they use, lie on the fact that the measurement is made without disturbing or influencing the MUT. Only when the sensor performs thermal ablation treatment, it is desired to influence the MUT. For the first goal of detecting the dielectric image some extraction mechanisms need to be developed. When simulations or measurements of the sensors are done, only the scattering parameters are obtained and they do not directly provide enough information to be able to extract straightforward the dielectric parameters. In this paper two different methods will be presented. First, the multi-resonant perturbation method (MRPM) that is able to extract the capacitive profile ($C_{LH,n}$) of a periodic structure influenced by a MUT. The second is a method based on network synthesis called the mixed Cauer expansion method (MCEM) that is able to extract the lumped element values of the equivalent circuit of the periodic structure as long as it can be modeled as a CRLH-TL.

For the accurate determination of the dielectric information of a MUT in the vicinity of a sensor, some key design objectives should be considered. The sensing element must be at an appropriate distance from the MUT and the sensitivity must be sufficient to ensure a good interaction. Additionally, the frequency must be low but without sacrificing accuracy in pursuance of cost reduction and to facilitate the needed technology. With this in mind the best suited structure is a metamaterial transmission line based sensor, with the advantages of having the sensing element in the series branch, being able to operate at a lower frequency and having higher sensitivity than a RH line with the same physical size. Also these types of structures are periodic and a detailed image can be obtained where the resolution of such an image is directly proportional to the amount of unit cells in the sensor. This brings additional challenges in the extraction since the only way is to use its equivalent circuit represented as a CRLH-TL as seen in Sect. 2.2. This means that in theory for each unit cell included in the sensor structure four parameters need to be extracted. This is not always the case, sometimes only the C_{LH} is enough to obtain the desired information.

M. Puentes Vargas, *Planar Metamaterial Based Microwave Sensor Arrays for Biomedical Analysis and Treatment*, Springer Theses,
DOI: 10.1007/978-3-319-06041-5_3,

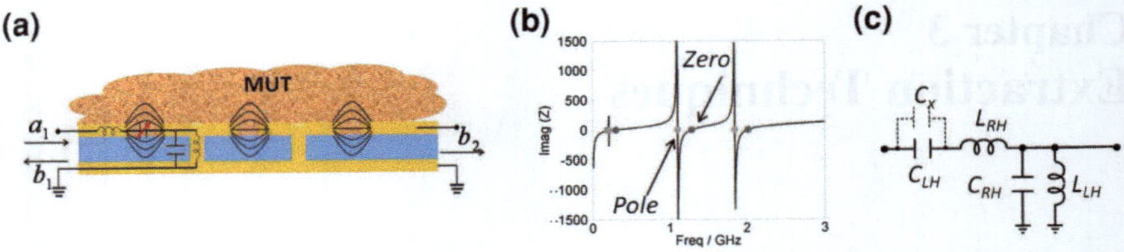

Fig. 3.1 Overview of the extraction process. **a** Sensor draft, **b** Input impedance plot and **c** Extracted equivalent circuit for one unit cell [1]

The MRPM and the MCEM are comparable in terms of what they can achieve. Both methods are able to extract a dielectric image of the MUT and therefore position of abnormalities can also be located. Both procedures have similar steps that are summarized in Fig. 3.1. In Fig. 3.1a is a schematic of what a planar sensor based on a CRLH-TL could look like with a MUT located on top of one of the unit cells for detection. It is a periodic structure with three unit cells and the LH elements are highlighted. The RH elements are included as parasitic effects as already discussed. In Fig. 3.1b the extracted imaginary part of the input impedance is depicted. This is the information used for the extraction of the dielectric information by applying either the MRPM or the MCEM. In the MRPM only the C_{LH} is extracted and with the MCEM all four parameters of each unit cell are extracted as seen in Fig. 3.1c. The capacitance C_x represents the influence of the MUT.

The MRPM was directly taken from the work presented in [2] and applied to the sensor structures realized in this work. On the other hand, the MCEM was developed within the framework of this thesis as a more accurate and complete alternative of the MRPM but unfortunately its application in real measurements prove to be quite difficult as will be seen in the next paper.

3.1 Multi-Resonant Perturbation Method

To understand how the MRPM behaves it is important to analyze first its simplest version the resonant perturbation method (RPM). In the RPM a MUT is inserted in a resonator, and the properties of the MUT are calculated from the change in the resonant frequency of the resonator due to the insertion of the MUT. The most widely used types of resonators are hollow metallic cavities and dielectric resonators. For the cavity perturbation, there are three types of RPM. The first method is the cavity-shape wall where part of the cavity wall is pulled or pushed out causing a change in the resonant frequency and the stored energy, but the energy dissipation in the cavity remains constant. The second method is the wall-impedance perturbation where a part of the cavity wall is replaced. It is often used for measuring the surface impedance of conductors. The final method is the material perturbation where a MUT is inserted in the cavity. The dielectric properties of the MUT can be measured

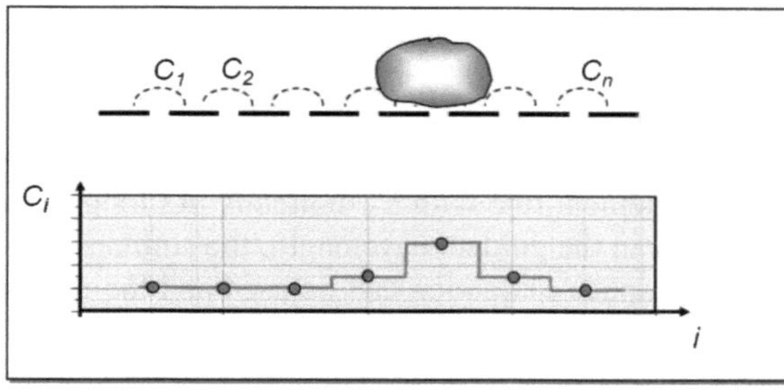

Fig. 3.2 Interaction between a CRLH-TL and loading perturbers and its correspondent capacitance profile [2]

with the shift in its resonance frequency. This method can be used in two scenarios. First, it works for the case when the complete resonator is filled with the original material and then is changed with the new material which has very similar dielectric properties to the original. Second, a small perturber is inserted in the resonator. It occupies a small portion of the resonator and the electromagnetic properties of the complete space except for the location of the perturber do not change. This method is based on three assumptions. First, the original medium in the cavity is lossless. Second, the perturbers are homogenous and are much smaller than the cavity, and finally the electromagnetic field outside the perturbers does not change [3]. This final case will be used in this work by combining it with metamaterial structures.

When considering a CRLH-TL with n_c unit cells as depicted in Fig. 3.1a, the lumped elements that compose its equivalent circuit can be used to calculate straightforward its line impedance Z_L as seen in Eq. 2.41 and its propagation constant β in Eq. 2.40. The input impedance when the line is open ended is given by

$$Z = jZ_L \cdot \cot(\beta d n_c), \tag{3.1}$$

where d is the unit cell length. The imaginary part of this input impedance has poles and zeroes when the electrical length of the line $\beta d n_c$ has multiples of $\lambda/4$ as depicted in Fig. 3.1b. The related frequencies are referred as resonance frequencies since the input impedance in these cases is zero or very high. When the line is loaded by perturbers as seen in Fig. 3.2, the resonance frequencies shift to lower values in dependance of the resonant mode and position of the perturber. The consideration of multiple resonant modes allows the simultaneous measurement of the dielectric properties of several perturbing objects by superposition of extracting changes in the LH capacitance of the equivalent circuit as can also be seen in Fig. 3.2.

The general formulation of the RPM relates the frequency shifts $\Delta\omega$ due to the loading of a MUT inside the resonator to the unloaded resonance frequency ω [3]

$$\frac{\Delta\omega}{\omega} = -\frac{\iiint_{V_c}(\Delta\varepsilon \cdot |E|^2 + \Delta\mu \cdot |H|^2)dV}{\iiint_{V_c}(\varepsilon \cdot |E|^2 + \mu \cdot |H|^2)dV}, \tag{3.2}$$

where V_c is the volume of the resonator filled with a material that has electromagnetic properties represented by ε and μ and E and H are the electric and magnetic fields respectively. The loading of the MUT is described by $\Delta\varepsilon$ and $\Delta\mu$. For the case of a one dimensional line resonator, Eq. 3.2 is reduced to an integration along a lumped element line with n_c unit cells. As long as the size of the unit cell remains smaller than the guided wavelength, the similarities between the wave equation and the telegrapher's equation can be used and the material parameters are replaced by its equivalent values in terms of the lumped elements for a CRLH-TL as follows

$$\varepsilon_{eff} = j\omega C_{RH} + \frac{1}{j\omega L_{LH}} \tag{3.3}$$

$$\mu_{eff} = j\omega L_{RH} + \frac{1}{j\omega C_{LH}}. \tag{3.4}$$

Then Eq. 3.2 can be rewritten:

$$\frac{\Delta\omega}{\omega} = -\frac{\sum_{i=1}^{n_c}\left[\Delta\left(\omega C_{RH} - \frac{1}{\omega L_{LH}}\right)\cdot|U_i|^2 + \Delta\left(\omega L_{RH} - \frac{1}{\omega C_{LH,i}}\right)\cdot|I_i|^2\right]}{\sum_{i=1}^{n_c}\left[\left(\omega C_{RH} - \frac{1}{\omega L_{LH}}\right)\cdot|U_i|^2 + \left(\omega L_{RH} - \frac{1}{\omega C_{LH,i}}\right)\cdot|I_i|^2\right]}. \tag{3.5}$$

Since the sensing element is the $C_{LH,p}$ in the pth unit cell, by loading the MUT only this capacitance will change and all other parameters will remain constant. From this assumption the numerator of Eq. 3.5 can be approximated to

$$\Delta\left(-\frac{1}{\omega C_{LH,p}}\right) = -\frac{1}{\omega(C_{LH} + \Delta C_{LH,p})} + \frac{1}{\omega C_{LH}} \approx \frac{\Delta C_{LH,p}}{\omega C_{LH}^2}. \tag{3.6}$$

The RPM assumes that only one capacitance is disturbed $C_{LH,p}$, therefore the sum in Eq. 3.5 can be extended to form the MRPM and the influence on several capacitances can be calculated by superposition. For this purpose, further simplifications will be made. At resonance the stored magnetic and electric energy are equal. Both energies are described in the denominator of Eq. 3.5. Hence, only one term with a factor of 2 needs to be considered. Taking into account this simplification plus what was stated in Eqs. 3.5 and 3.6 can be further simplified into:

$$\frac{\Delta\omega}{\omega} = -\frac{\frac{1}{\omega C_{LH}}}{2\cdot\left(\omega L_{RH} - \frac{1}{\omega C_{LH}}\right)}\frac{\frac{\Delta C_{LH,p}}{C_{LH}^2}\cdot|I_p|^2}{\sum_{i=1}^{n_c}|I_i|^2}, \tag{3.7}$$

with the resonance condition for the line given as:

$$\beta \cdot n_c \cdot d = -n \cdot \frac{\pi}{2} \text{ with } n = 1, 2, 3... \tag{3.8}$$

These resonances are related to the zeroes and poles of the input impedance. For an open ended line, $n = 1, 3, 5...$ is related to zeroes and $n = 2, 4, 6...$ to poles of the input impedance. With the assumption of a sinusoidal current distribution, the current through the pth capacitor is

$$I_p = I_0 \cdot \sin(\beta(p-1)d). \tag{3.9}$$

Then Eq. 3.7 can be developed for the nth resonance to

$$\frac{\Delta\omega_n}{\omega_n} = -\frac{\frac{1}{\omega C_{LH}}}{2 \cdot \left(\omega L_{RH} - \frac{1}{\omega C_{LH}}\right)} \cdot \frac{\left|\sin\left(\frac{n}{n_c}\frac{\pi}{2}(p-1)\right)\right|^2}{\sum_{i=1}^{n_c}\left|\sin\left(\frac{n}{n_c}\frac{\pi}{2}(i-1)\right)\right|^2} \cdot \frac{\Delta C_{LH,p}}{C_{LH}^2}, \tag{3.10}$$

and the case of more than one perturber can be then calculated by superposition:

$$\frac{\Delta\omega_n}{\omega_n} = \sum_{p=1}^{n_c} a_{n,p} \cdot \frac{\Delta C_{LH,p}}{C_{LH}}, \tag{3.11}$$

with

$$a_{n,p} = -\frac{\frac{1}{\omega C_{LH}}}{2 \cdot \left(\omega L_{RH} - \frac{1}{\omega C_{LH}}\right)} \cdot \frac{\left|\sin\left(\frac{n}{n_c}\frac{\pi}{2}(p-1)\right)\right|^2}{\sum_{i=1}^{n_c}\left|\sin\left(\frac{n}{n_c}\frac{\pi}{2}(i-1)\right)\right|^2}. \tag{3.12}$$

In Fig. 3.3 is drawn a cylindrical resonator with a perturber inside. The interaction between the perturber and the electric field for two different resonance frequencies is depicted. At $\Delta\omega_4$ the perturber is located in a zero value of the field and at $\Delta\omega_{16}$ is located in the maxima of the field. From this figure and Eq. 3.11 can be seen that a change on one capacitance $\Delta C_{LH,p}$ has an impact in all resonance frequencies which corresponds to a shift in the poles and zeroes of the input impedance. Since in the RPM only small changes are considered, their influence is linearized in Eq. 3.6.

As a result an equation system can be found that relates the left-handed capacitance changes to the changes in the resonance frequencies of the line. To determine n_c unknown capacitances is required the same amount of equations as follows:

$$\begin{pmatrix} \Delta\omega_1 \\ \Delta\omega_2 \\ \Delta\omega_3 \\ \cdots \end{pmatrix} = \begin{pmatrix} a_{1,1} & a_{1,2} & a_{1,3} & \cdots \\ a_{2,1} & a_{2,2} & a_{2,3} & \cdots \\ a_{3,1} & a_{3,2} & a_{3,3} & \cdots \\ \vdots & \vdots & \vdots & \ddots \end{pmatrix} \begin{pmatrix} \Delta C_1 \\ \Delta C_2 \\ \Delta C_3 \\ \cdots \end{pmatrix} \tag{3.13}$$

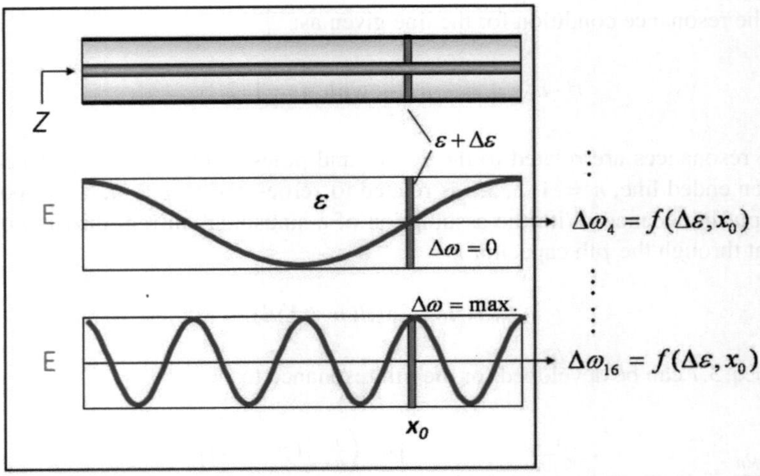

Fig. 3.3 Changes produced on the resonance frequencies by a perturber located inside a resonator

$$\Delta\omega = A \cdot \Delta C_{LH}, \tag{3.14}$$

where the vector ΔC_{LH} represents the relative changes in LH capacitances of the transmission line and the vector $\Delta\omega$ corresponds to the relative resonance frequency shifts. If the transmission line is build up of n_C cells, both vectors will consist of n_C elements and hence, the dimension of the matrix A is $n_C \times n_C$. The construction of matrix A is the biggest disadvantage of this method. It requires $n_c + 1$ calibration measurements and for example, for the sensors presented in Chap. 5 where microfluidic channels are included, the procedure can be time consuming and difficult. On the other hand the matrix has to be inverted A^{-1} to obtain the capacitive profile ΔC_{LH} from the relative resonance frequency shifts $\Delta\omega$ and sometimes the obtained matrix is no adequate to make the inversion. In that particular case the method cannot be used to make the extraction.

The CRLH-TL with $4 \cdot n_c$ unit cells have $2 \cdot n_c$ poles and zeroes in the LH-band and RH-band respectively. Since only n_c values are necessary for the extraction of n_C capacitances, they can be extracted only from the LH-band which is significantly lower in frequency and smaller in bandwidth [2]. This gives the method an advantage since only the measurement of the LH-band is necessary.

The principle of operation of this method can be summarized as first making the calibration measurements by loading each sensor unit cell with a known material for constructing the A matrix. To minimize errors in the extraction, the calibrating material should have a dielectric constant as close as possible to the MUT. Once this is done the sensor can be loaded with any MUT. Finally the MRPM is applied and the C_{LH} are calculated. Several examples of extracted capacitance profiles from real measurements using a CRLH-TL sensor with 9 unit cells can be seen in Fig. 3.4. The profiles include loading on several unit cells and sometimes on neighboring unit cells

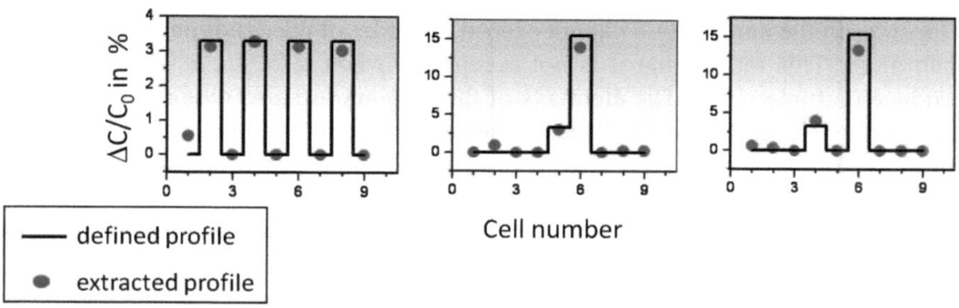

Fig. 3.4 Several examples of extracted capacitance profiles from real measurements of a CRLH-TL sensor developed within the framework of this work [2]

to test if a one unit cell resolution can be achieved. The small errors that sometime occur are due to the linearization of the system of equations that sometimes can introduce approximation errors. How big the errors are is dependant on the available resolution for the frequency shifts extraction as well as the amount of unit cells and permittivity difference between the calibration material and the actual MUT. Nevertheless, the extracted profiles have a very good agreement with the defined profile making the MRPM an excellent tool for extracting the dielectric properties of MUT in the vicinity of a CRLH-TL.

3.2 Mixed Cauer Expansion Method

For most networks it is possible to determine its response by knowing its topology and source of excitation. But when the topology is unknown but the source of excitation and its response are known, it is possible to use network synthesis to determine the topology. Nevertheless, network synthesis does not provide a unique solution when the network is causal and stable. By causal, it is understood that the network cannot have a response in absence of an excitation signal. This can be written as $h(t) = 0$ for all $t < 0$. Where $h(t)$ is the time response of a given network system. By stable, it is understood that for a bounded input there should be a bounded output. When the function is transformed in the frequency domain by means of the Laplace transform, this can be represented by the fact that all the poles of the system must lie in the left plane and they are simple. A frequency response $H(s)$ of a system looks as follows

$$H(s) = \mathcal{L}\{h(t)\}(s) = \int_0^{\infty} h(t)\, dt \tag{3.15}$$

$$H(s) = \frac{N(s)}{D(s)} = \frac{a_n s^n + a_{n-1}s^{n-1} + \cdots + a_1 s + a_0}{b_m s^m + b_{m-1}s^{m-1} + \cdots + b_1 s + b_0}. \tag{3.16}$$

The order of the numerator n cannot exceed the order of the denominator m by more than unity. This means that it is not acceptable when $n > (m + 1)$ since it will produce multiple poles. The elements of the network are also required to be linear and time-invariant. To satisfy all these requirements the polynomials of Eq. 3.16 must fulfill the Hurwitz condition. This means they must be real when s is real and the real parts of the roots of the polynomial should be zero or negative. Additionally the Hurwitz polynomials must have the following properties. The coefficients of s must be positive, the roots must lie in the left plane or $j\omega$ axis and all the quotients during a continued fraction expansion must be positive.

The input impedance or admittance of a network system can be represented by the previous $H(s)$ as long as all the mentioned properties are fulfilled. Depending on the topology, there exist different methods to do the network synthesis. The most common ones from the literature are the Foster or the Cauer canonical form methods [4]. The basic philosophy of the synthesis is that for a given positive real function, for example the impedance of a network Z, it can be decomposed into smaller functions and each of them must also be positive and real. Each function will represent a single extracted element R, L or C or a suitable parallel combination of them. For actually decomposing the main function four different mechanisms can be used. They are the removal of a pole at infinity, removal of a pole at origin, removal of a conjugated imaginary pair of poles and removal of a constant. A combination of several of this mechanisms is also possible [4].

The topologies relevant to the sensor addressed in this work are the LH-TL and RH-TL. For these cases the cauer canonical form (CCF) I and II are appropriate. When an ideal LH-line is considered as shown on Fig. 2.2b, the equivalent circuit can be extracted with the Cauer canonical form (CCF) II where the impedance transfer function is assumed to have the degree of the numerator $N(s)$ lower than that of the denominator $D(s)$ and must differ by unity. Otherwise, we must consider the admittance $Y(s) = 1/Z(s)$ instead of $Z(s)$. The coefficients of both polynomials must be arranged in ascending power of s and a continued fraction expansion is then performed:

$$
\begin{aligned}
Z(s) &= \frac{V_{in}}{I_{in}} = \frac{N(s)}{D(s)} = \frac{a_0 + a_2 s^2 + a_4 s^4 + \cdots + a_k s^k}{b_1 s + b_3 s^3 + b_5 s^5 + \cdots + b_{k+1} s^{k+1}} \\
&= \frac{1}{C_1 s} + \frac{1}{\frac{1}{L_2 s} + \frac{1}{\frac{1}{C_3 s} + \frac{1}{\frac{1}{L_4 s} + \frac{1}{\cdots}}}}.
\end{aligned}
\tag{3.17}
$$

From Eq. 3.17 the values of each inductor and capacitor of the equivalent circuit can be calculated and it is clear that on each step of the continued fraction expansion a pole at the origin is removed. With this method only the calculations for the ideal LH-line can be performed with very high accuracy but for the real case where parasitic effects are considered, e.g. in a more complicated structure such as the CRLH-TL, a combination of the CCF I and II has to be implemented. The CCF I has a behavior very similar to the CCF II, the differences lie in the following aspects: The equivalent circuit is a RH-line as shown on Fig. 2.2a, the coefficients of both polynomials must be arranged in descending power of s, the degree of the numerator $N(s)$ must be higher than that of the denominator $D(s)$ and finally on each step of the continued fraction expansion a pole at infinity is removed as presented on the following equation:

$$\begin{aligned} Z(s) &= \frac{N(s)}{D(s)} = \frac{a_k s^k + a_{k-2} s^{k-2} + \cdots + a_3 s^3 + a_1 s}{b_{k-1} s^{k-1} + b_{k-3} s^{k-3} + \cdots + b_2 s^2 + b_0} \\ &= L_1 s + \cfrac{1}{C_2 s + \cfrac{1}{L_3 s + \cfrac{1}{C_4 s + \cfrac{1}{\cdots}}}}. \end{aligned} \tag{3.18}$$

When the two CCF are used in an alternating fashion, it is possible to extract a CRLH equivalent circuit as shown in Fig. 2.2c. This method is called the mixed Cauer expansion method (MCEM) and was developed within the framework of this work. The starting transfer function must be arranged in descending power of s because the first element to be extracted is L_{RH}, this means a CCF I is performed and a pole at infinity is removed. Then the polynomials must be arranged in ascending power of s and inverted to extract C_{LH} with the CCF II and remove a pole at origin. The third step is to arrange the polynomials in descending power of s again and perform a CCF I to extract C_{RH}. The last step consists in arranging the polynomials in ascending power of s and inverting them one last time for the CCF II and the extraction of L_{LH}. This procedure completes the extraction of one CRLH unit cell and can be repeated for several unit cells as necessary according to the size of the actual structure. A diagram of the steps of the procedure is shown in Fig. 3.5. The expanded impedance transfer function will look as follow:

$$Z(s) = L_{RH} s + \frac{1}{C_{LH} s} + \frac{1}{C_{RH} s + \frac{1}{L_{LH} s}} + \cdots \tag{3.19}$$

In summary, the principle of operation can be explained when considering the sensor element with no MUT in the vicinity, then from the measurement of the reflection coefficient the input impedance of the open ended line at a certain frequency can be extracted. From this impedance information the poles and the zeros are found and the transfer function of Z is then constructed. By using the MCEM all capacitance and inductance values of the CRLH unit cells can be determined. After obtaining these reference values, the measurement with MUT in the vicinity of the sensor can be done and the same procedure is performed. For the detection of the position of each object, it is only necessary to compare these values with the ones obtained in the reference measurements. To prove that the algorithm based on the MCEM is appropriate, a comparison between the original values and the extracted values for a simulated structure with two identical CRLH unit cells is presented in Fig. 3.6. In Fig. 3.6a are the extracted capacitances C_{LH} and C_{RH} and in Fig. 3.6b are the inductances L_{LH} and L_{RH}. The obtained accuracy is satisfactory as the capacitances that are the most important parameter we need are almost equal to the defined profile.

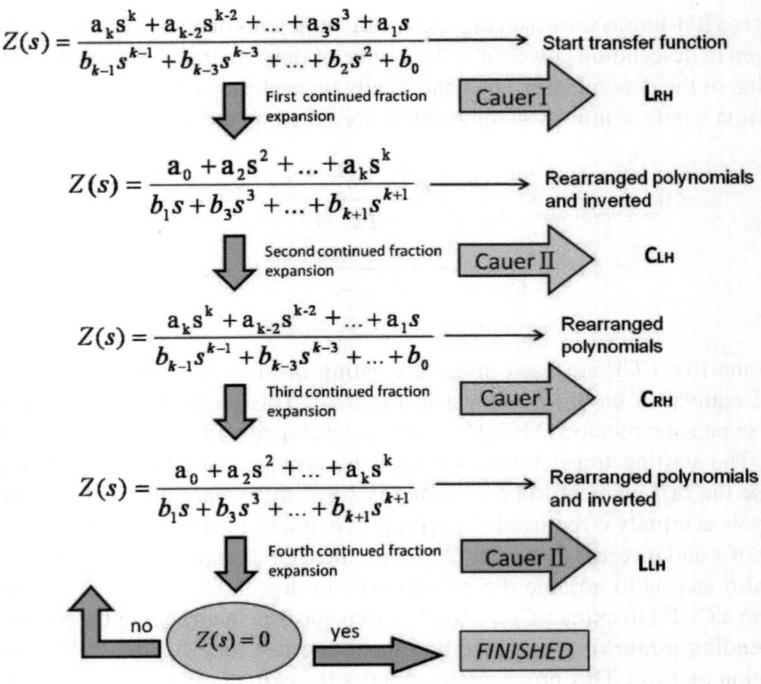

Fig. 3.5 Diagram representing the steps of the continued fraction expansion of the mixed Cauer expansion method (MCEM)

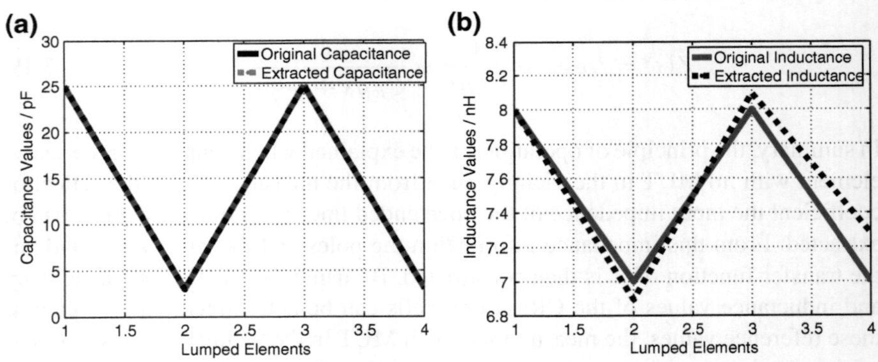

Fig. 3.6 Extracted **a** capacitances C_{LH} and C_{RH} and **b** inductances L_{LH} and L_{RH} of a simulated structure with two identical unit cells using the MCEM [1]

It is also important to emphasize that with this approach a one cell resolution can be achieved and perturbers located next to each other can be resolved where the performance is limited only by the signal to noise ratio. A small SNR causes big difficulties in extracting the poles and the zeroes resulting in an erroneous impedance transfer function.

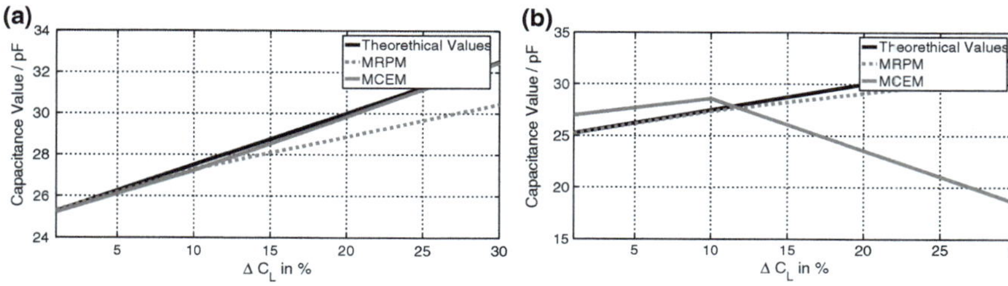

Fig. 3.7 Comparison between the MRPM and the MCEM extraction methods for **a** three unit cells and **b** five unit cells [1]

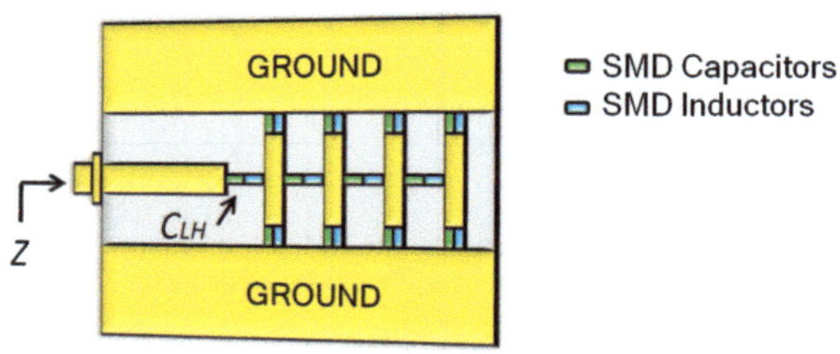

Fig. 3.8 Coplanar simulated structure with four unit cells [1]

3.3 Comparison of the Extraction Techniques

To compare the two presented methods a CRLH structure made of lumped elements was simulated with three identical unit cells. The C_{LH} of the middle unit cell was changed 1, 10 and 30 %. The MRPM and the MCEM where applied and the results are shown in Fig. 3.7a. For this particular example, the MCEM presents a better accuracy and gives information about all CRLH lumped elements which is a nice bonus feature but in additional tests where the amount of unit cells was increased to five unit cells the results are different. In this case the MCEM presents a big numerical error due to the residual errors produced in the continued fraction expansion. For this case the MRPM maintains a better accuracy as shown in Fig. 3.7b.

To further prove the performance of the two methods a left handed coplanar design with 4 unit cells was simulated using the full-wave simulator CST Microwave Studio. The structure is depicted in Fig. 3.8. SMD elements where used to better control the values of the lumped elements in the CRLH structure and reduce the frequency of operation. Different capacitive profiles where constructed where the C_L was changed in the range of 1–20 %. The first scenario was to alter the C_L in the second unit cell by $1, 5, 10$ and 20 %. The results are shown on Fig. 3.9 for both methods. From the four presented profiles it can be seen once more that the MCEM has accuracy problems due to the method itself where continued divisions of the polynomials are performed. This is only aggravated in a real case scenario where organic material is tested and its high losses will shift the imaginary part of the impedance making

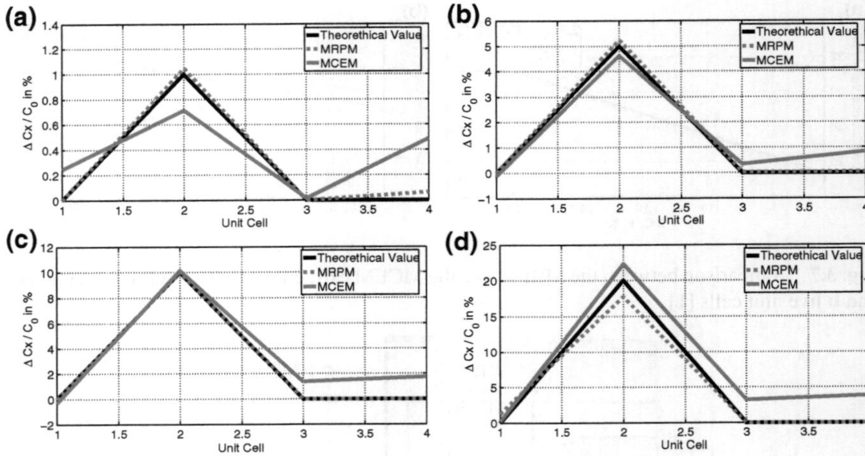

Fig. 3.9 Comparison between the MRPT and the MCEM. **a** Capacitance change of 1 %. **b** Capacitance change of 5 %. **c** Capacitance change of 10 % and **d** Capacitance change of 20 % [1]

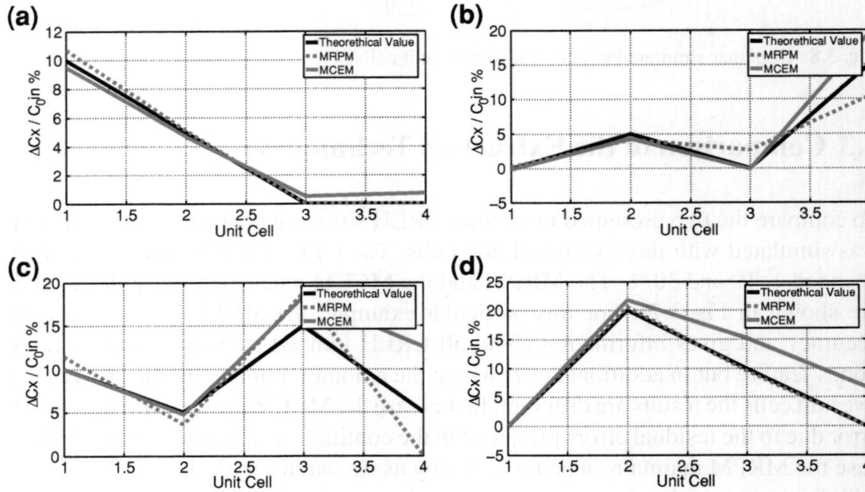

Fig. 3.10 Comparison between the MRPT and the MCEM with different capacitive profiles [1]

it very difficult to construct an appropriate impedance transfer function. In future work further signal processing mechanisms such as optimization techniques can be implemented to improve the method since it shows great promise with its biggest advantage being that it does not need tedious calibration measurements. Nevertheless for this example the MRPM shows higher performance and accurate results.

The second scenario was to create different capacitive profiles where more than one unit cell was altered in order to test the one unit cell resolution of the methods. The results are shown in Fig. 3.10. From the results it is clear that the capacitive profile based on the relative change of the capacitances can be detected with both methods and the extraction is done with a one unit cell resolution even when the changes are small. In the case of the MCEM the extraction is adequate but the previously mentioned problems of this method can still be seen. On the other hand, the MRPM shows also a good approximation of the profiles. The errors are again due to the linearization inherent from this method. Although the method has the disadvantage of the matrix A, it seems to behave better than the MCEM for real measurement scenarios and it is anyway a simple, but very powerful approach. Additionally it has the advantage of needing only information of one (RH or LH) band since only one forth of the poles and zeros need to be detected which can result in a cost reduction of the sensor. Both methods will be used within the following papers for different sensing scenarios.

References

1. M. Puentes, M. Schüßler, C. Damm, R. Jakoby, Extraction of capacitive profiles with CRLH transmission lines. Frequenz **64**(5/6), 74–78 (2010)
2. M. Schüßler, M. Puentes, C. Mandel, R. Jakoby, Multi-resonant perturbation method for capacitive sensing with composite right-left handed transmission line, in *Proceedings of IEEE MTT-S International Microwave Symposium Digest*, (Anaheim, June 2010)
3. L.F. Chen, C.K. Ong, C.P. Neo, V.V. Varadan, V.V.K. Valev, *Microwave Electronics* (Wiley, New York, 2004)
4. S. Ghosh, *Network Theory, Analysis and Synthesis* (Prentice Hall, New Delhi, 2005)

Chapter 4
Basic Sensor Concept for the Detection of a Spatial Material Distribution

4.1 Introduction of the Basic Sensing Principle

The modern industry is interested in new universal adaptable and affordable sensor concepts. These sensor concepts should be suitable for monitoring or control systems in a wide range of applications. For example, the accurate determination of the materials concentration, properties and their velocity in industrial conveyor belt- and chute systems. Although these structures are at least on one side open, the same sensor principles like in pipelines and tubes can be applied [1, 2]. These principles can rely on mechanical-, acoustic-, electrostatic and electromagnetic effects with all their specific advantages and disadvantages. The work in this area started, in the Institute of Microwave Engineering and Photonics work in this area started with a cylindrically shaped mass flow sensor for particulate solids [3]. The central sensor element is an adapted metamaterial-transmission line resonator operated in a high harmonic mode. The sensor element output signal is the phase variation in the transmitted RF signal originating from moved material in the vicinity of the resonator. By the skillful application of signal processing methods the dielectric constant and velocity of a transported material can be detected in real time. A test facility was built for measurements to prove the concept.

In this work is studied the basic principle for a planar metamaterial sensor aimed for biomedical applications. The sensors are able to detect the spatial distribution and relative dielectric properties of the material under test (MUT). Different geometries are tested and the extraction methods presented in Chap. 3 are applied for the information extraction. The principle of operation is depicted in Fig. 4.1 and consists of a planar metamaterial sensor that comes into direct contact with the MUT. The sensor is a CRLH-TL and the C_{LH} is located in the gap between the patches. An excitation signal is inserted and the electric field is able to interact with the MUT depending on the design of the sensor and the dielectric properties of the MUT. From the reflected and/or transmitted signal the information about the dielectric properties of the MUT and its distribution within the sensor array can be derived through the extraction of a capacitive profile. This principle can be customized for any application where the

M. Puentes Vargas, *Planar Metamaterial Based Microwave Sensor Arrays for Biomedical Analysis and Treatment*, Springer Theses,
DOI: 10.1007/978-3-319-06041-5_4,

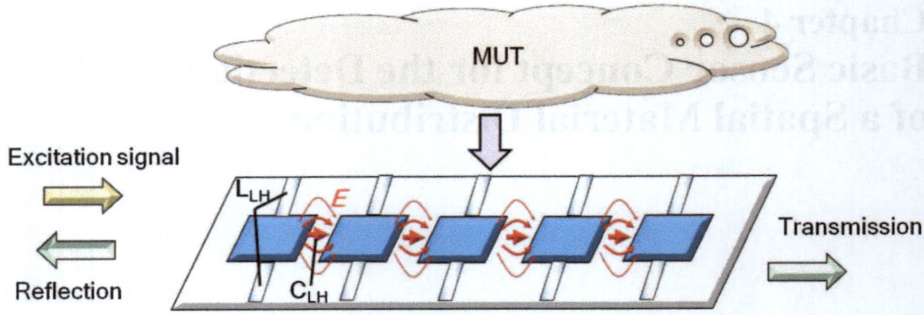

Fig. 4.1 Principle of operation for a planar metamaterial sensor. From the reflected and/or transmitted signal the information about the dielectric properties of the MUT and its distribution within the sensor array can be derived through the extraction of a capacitive profile

dielectric properties of a material are of importance. That includes environmental monitoring, industrial applications, biomedical applications, and many more.

The application chosen to make the initial tests of the presented principle is for the paper industry where the quality of raw paper can be monitored non-invasively with these sensors. The production process from the raw material wood to the complete paper has several steps. After conducting several manufacturing steps like the bleaching, blending and felting the paper web is ready to be processed. At this point of the production chain, it is essential to monitor the product for manufacturing errors. This monitoring step also has to be repeated after pressing and drying the paper web. For this application microwave sensors are suitable. Although these sensors are generally more expensive than sensors working at lower frequencies, they feature properties that make the cost worthwhile. Specially for moveable environments, which can be found in many industrial sites, microwave sensors are advantageous, since they provide a contact-less and non-invasive sensing technique. On the other hand, the interaction between MUTs and electromagnetic waves is well known for giving valuable information at a wide range of frequencies. Nevertheless, there are special spectral signatures only present in the microwave range. These properties of the MUT at microwave frequencies can be exploited to analyze structure, function or composition of the MUT.

4.2 Composite Right/Left Handed Transmission Line Used for Microwave Sensors

CRLH-TL offer several degrees of freedom in their design to feature very different properties. Depending on the application at hand, they can be exploited to design different sensor structures. As explained in Sect. 2.1, CRLH-TL include a purely left-handed part as well as a purely right-handed made of lumped element as seen in Fig. 2.2. It is interesting to analyze what advantages can bring the use of LH structures

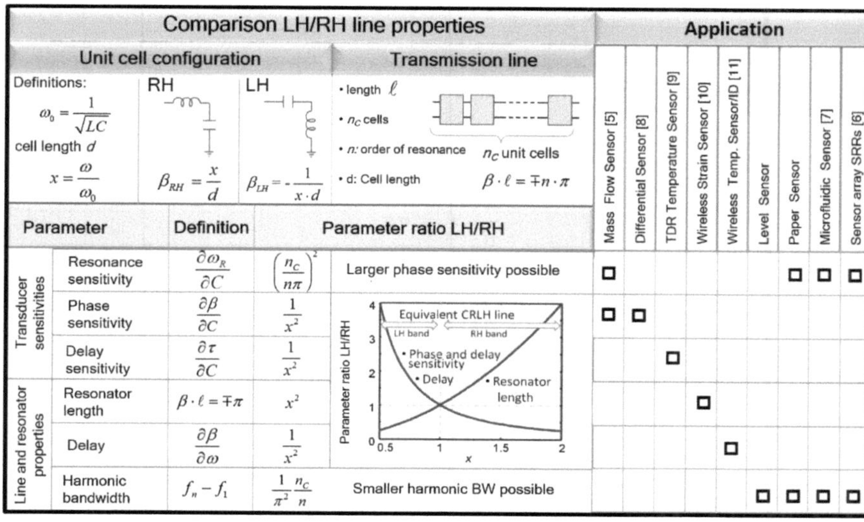

Fig. 4.2 Overview of sensor relevant transmission line properties and their possible applications [4]

for sensor concepts. For this purpose a list of the relevant line properties for sensor design is depicted in Fig. 4.2 and its values are normalized to the properties of the corresponding RH-TL to understand the advantages of the LH-TL part. In the upper part of the figure are again given the equivalent circuits of the two cases that includes only reactive elements with the same nominal value. All relevant definitions such as characteristic frequency ω_0, cell length d, propagation constant β, among others are also included. Since some of the parameters that are analyzed to compare the two cases are frequency dependant, the characteristic frequency ω_0 is very important, specially, the ratio x of the operation frequency ω to ω_0. The CRLH line can be approximated with LH behavior for x being less than unity and RH behavior for x being larger than unity.

The first set of parameters entitled "Transducer sensitivities" is related to the properties of a CRLH-TL as a transducing element since they describe the influence of a change in the capacitance due to an electromagnetic property of the line such as time delay, characteristic frequency or phase. The change in these line properties can be monitored by a dedicated electrical circuit, and hence, allows the extraction of information on the MUT. Lets consider the first parameter from this group, resonance sensitivity. This parameter is very important for most of the applications that will be presented in this work, since their principle of operation is based on tracking changes in this resonance frequencies due to the interaction between the sensor and the MUT [5–7]. For resonance condition the electrical length must have an integer multiple of π as seen in the following equation

$$\beta \cdot \ell \overset{!}{=} \pm n\pi \qquad n \in \mathbb{N} \tag{4.1}$$

The RH-TL resonance frequency is defined as

$$\beta \cdot \ell = n\pi \tag{4.2}$$

$$\frac{x}{d} \cdot \ell = n\pi \tag{4.3}$$

$$\omega_{R,RH} = \frac{n\pi\omega_0}{n_c}, \tag{4.4}$$

and the LH-TL resonance frequency is defined as

$$\beta \cdot \ell = -n\pi \tag{4.5}$$

$$-\frac{1}{x \cdot d} \cdot \ell = -n\pi \tag{4.6}$$

$$\omega_{R,LH} = \frac{\omega_0 n_c}{n\pi}. \tag{4.7}$$

To revise the change in resonance frequency due to a capacitance change of the unit cell we derive Eqs. 4.4 and 4.7 as follows

$$\frac{\partial\omega_{R,RH}}{\partial C} = \frac{-n\pi}{2n_c C\sqrt{LC}} \tag{4.8}$$

$$\frac{\partial\omega_{R,LH}}{\partial C} = \frac{-n_c}{2n\pi C\sqrt{LC}}, \tag{4.9}$$

and then the ratio between the LH-TL resonance frequency and the RH-TL resonance frequency is calculated as

$$\frac{\partial\omega_{R,LH}/\partial C}{\partial\omega_{R,RH}/\partial C} = \left(\frac{n_c}{n\pi}\right)^2. \tag{4.10}$$

By analyzing this result and considering for example a structure with 10 unit cells and order of resonance $n = 1$ the LH-TL can have 10 times higher sensitivity than the corresponding RH-TL.

The second parameter, phase sensitivity, is important for sensors where the tracking of changes in the phase is the main operation principle. Examples developed within the IMP research group are the mass flow sensor for monitoring of coal in pipeline applications [5] and the differential sensor for monitoring a MUT with very high sensitive since it corrects any fabrication tolerances or temperature influence within the same structure due to its operation principle where a reference line and a

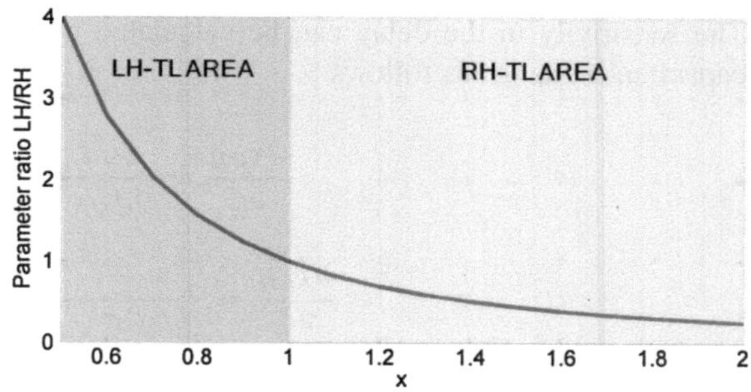

Fig. 4.3 Comparison of the phase sensitivity, delay sensitivity and delay between the LH-TL and the RH-TL

measurement line are connected in a single output [8]. The change in the phase due to a change in the capacitance of the unit cell can be expressed for both RH-TL and LH-TL as

$$\frac{\partial \beta_{RH}}{\partial C} = \frac{\omega \sqrt{L}}{2d\sqrt{C}} \tag{4.11}$$

$$\frac{\partial \beta_{LH}}{\partial C} = \frac{1}{\omega d C \sqrt{LC}}, \tag{4.12}$$

and the ratio between Eqs. 4.12 and 4.11 is

$$\frac{\partial \beta_{LH}/\partial C}{\partial \beta_{RH}/\partial C} = \frac{\omega_0^2}{\omega^2} = \frac{1}{x^2}. \tag{4.13}$$

The result can be seen in Fig. 4.3. The areas where the LH-TL can have a higher phase sensitivity are highlighted and up to 4 times larger values can be obtained if the operation frequency is chosen close to the cut-off frequency of the LH-TL.

The third parameter is delay sensitivity. It is useful for wireless sensors, an example is the time domain reflectometry (TDR) temperature sensor presented in [9]. The delay of a transmission line is given by

$$\tau = \frac{\partial \varphi}{\partial \omega} = \frac{\partial \beta \cdot \ell}{\partial \omega}, \tag{4.14}$$

and the RH-TL and LH-TL have a delay represented by

$$\tau_{RH} = \frac{\ell}{\omega_0 d} \tag{4.15}$$

$$\tau_{LH} = \frac{\ell \omega_0}{d\omega^2}. \tag{4.16}$$

The sensitivity in the delay can be calculated by a partial derivative in terms of capacitance change as follows

$$\frac{\partial \tau_{RH}}{\partial C} = \frac{\ell\sqrt{L}}{2d\sqrt{C}} \tag{4.17}$$

$$\frac{\partial \tau_{LH}}{\partial C} = \frac{-\ell}{2d\omega^2 C\sqrt{LC}}. \tag{4.18}$$

Finally, the ratio between both delay sensitivities is given by

$$\frac{\partial \tau_{LH}/\partial C}{\partial \tau_{RH}/\partial C} = \frac{\omega_0^2}{\omega^2} = \frac{1}{x^2}. \tag{4.19}$$

The result has a similar behavior as the phase sensitivity and can be analyzed also with Fig. 4.3. Again by proper selection of the operation frequency a higher delay sensitivity can be obtained by operating the sensor in the LH-band.

The second set of parameters are "Line and resonator properties" at the lower left of Fig. 4.2, which are more classical circuit and system design parameters as resonator size, harmonic bandwidth and delay. These parameters are decisive for the practical usage of a CRLH-TL as a sensor. Beside the sensitivity properties of the first set of parameters, the second set of parameters enable miniaturization and the reduction of the required bandwidth. In this category the first parameter is resonator length. When LH-TL properties are compared to homogenous, distributed element lines as microstrip or coplanar line structures, the sensitivity of the LH-TL will be dramatically larger since homogenous lines have smaller length-related values of capacitances and inductances. This statement is also applicable to the length of transmission line resonators, meaning that LH resonators can have extremely small dimensions. This can be seen by comparing the resonator length between an LH-TL and an RH-TL as follows

$$\beta_{RH} \cdot \ell = \pi \tag{4.20}$$

$$\beta_{LH} \cdot \ell = -\pi \tag{4.21}$$

$$\frac{\ell_{LH}}{\ell_{RH}} = \frac{-\pi/\beta_{LH}}{\pi/\beta_{RH}} = -\frac{x/d}{-1/xd} = x^2. \tag{4.22}$$

The result can be seen in Fig. 4.4 and the area where the LH-TL can have a shorter resonator length in comparison with the RH-TL is highlighted. An example where the resonator length was a decisive parameter in the design can be found in [10]. The presented concept is a wireless strain sensor with identification which uses microstrip patch antennas, miniaturized by the use of a CRLH-TL, for passive frequency domain identification generation.

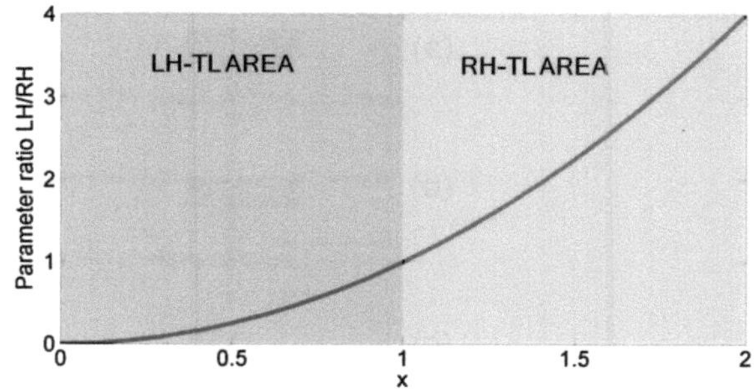

Fig. 4.4 Comparison of the resonator length between the LH-TL and the RH-TL

The next parameter, time delay, shows a similar behavior as the phase sensitivity and delay sensitivity. An example of how this parameter is important in the design of sensors can be seen in [11], where a passive and chipless wireless sensor tag with identification is depicted. Time delay was already defined in Eq. 4.14 and for both cases it was shown in Eqs. 4.15 and 4.16. The ratio is then defined by

$$\frac{\tau_{LH}}{\tau_{RH}} = \frac{\frac{\ell\omega_0}{d\omega^2}}{\frac{\ell}{\omega_0 d}} = \frac{1}{x^2}. \tag{4.23}$$

The result was shown in Fig. 4.3. Once more the adequate selection of the operation frequency can impact the performance of the sensor, in this case the desired time delay can be controlled. The last parameter in the list is the harmonic bandwidth which is rarely used as a measure, since a large number of sensors work only at one operation frequency. Nevertheless, it can be used for specific applications and it is defined as

$$BW = \omega_R(n) - \omega_R(1). \tag{4.24}$$

For the RH-TL it takes the form

$$BW_{RH} = \frac{n_c(\omega_0 - n\Omega_0)}{n\pi}, \tag{4.25}$$

and for the LH-TL

$$BW_{LH} = \frac{\pi(n\omega_0 - \Omega_0)}{n_c}. \tag{4.26}$$

The ratio is then calculated as

$$\frac{BW_{LH}}{BW_{RH}} = \frac{1}{\pi^2}\frac{n_c}{n}. \tag{4.27}$$

In the case of the LH-TL, the required bandwidth is reduced by a factor of approximately $1/10$ if all resonances are used ($n = n_c$). In this work some sensor applications

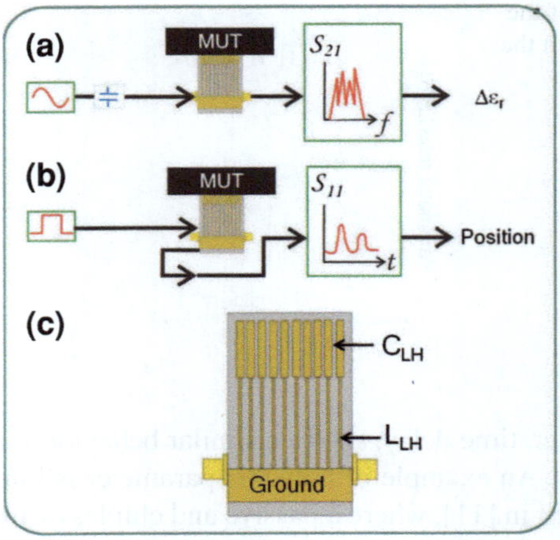

Fig. 4.5 Block diagram of the complete sensor setup: **a** Resonant mode of operation, **b** Transmission line mode of operation and **c** Design of the sensor structure [12]

will be presented that use multiple harmonic resonances for the measurement of spatially resolved data. Hence, the required bandwidth, meaning the absolute difference between the fundamental and a higher resonance could be a very important figure of merit. A summary of all the listed properties in graphical form is also included in the Fig. 4.2 as well as possible sensor concepts in the right side of the figure, some of which are described in detail in this and the following chapters. It should be highlighted that the inductors can also be used to influence the line properties as key components of the CRLH-TLs. The comparison made of the different parameters is also valid in the case the inductors are used to manipulate the line. The reason for not using them in the design of the sensors presented in this work is because the main aim is to develop planar sensors. Since the capacitor is located in the serial branch for the LH-TL, it can be straightforward used as transducer element. The inductor, on the other hand, is located in the parallel branch and it would be cumbersome to locate it near the MUT. A final advantage of using CRLH-TLs for sensor applications is the extreme geometrical flexibility that they provide. Since they are basically lumped element circuits, they are not limited to a PCB based implementation as microstrip or other TLs. The possibility to have 3D setups with wired lumped elements behaving like TLs offers the flexibility to bring the sensitive sensor area as close as possible to the most suitable position.

4.3 Planar Microwave Sensor

The first developed sensor has as a central sensing element, an adapted planar metamaterial-transmission line that can be operated in two modes. Figure 4.5 displays the block diagram of the complete sensor system. On part (a) of the figure is shown one operation mode where the line is used as a resonator in a high harmonic mode. It can be used to measure relative changes in the dielectric constant of the MUT on the sensor element by tracking changes in the phase of the structure. Since the sensor operates in a resonant mode with sharp resonance frequencies, the phase change will be more pronounced and the detection will be more accurate. By changing the coupling to the sensor element from capacitive to galvanic as shown on part (b) of the figure, the sensor element can be used as a transmission line. In this operation mode, time domain reflectometry techniques are employed to determine the spatial material distribution on the element or inhomogeneities in the MUT. The sensor setup consists of the sensor element, a test facility for measurements with the MUT, and the detection and post-processing of the resulting signal. As for the sensor itself the design is unconventional. It is a planar structure with a very special geometry, where both, the sensor element and the ground plane are on the same level. The feeding is done through a TL that encounters the first unit cell made of a long thin line to represent the L_{LH} and a patch element to represent the C_{LH}. The RH components are of course present as parasitic effects. Then, a periodic arrangement of the initial unit cell is done to form a structure with 9 unit cells as seen in Fig. 4.5c.

4.3.1 Resonant Mode of Operation

In the first mode of operation of the proposed sensor the relative changes in the dielectric constant of a MUT can be detected. The network topology can be straightforwardly derived from the theory of planar metamaterial transmission lines presented in Sect. 2.2. The structure consists of a periodical CRLH-TL as seen in Fig. 2.2 but for the analysis of this mode of operation lets consider the RH and LH components separately. When considering an RH unit cell or an LH unit cell in the balanced case the transition frequency ω_0 is defined as

$$\omega_0 = \omega_s = \omega_p = \frac{1}{\sqrt{L'_{RH} C'_{LH}}} = \frac{1}{\sqrt{L'_{LH} C'_{RH}}}, \tag{4.28}$$

and β for both cases was already defined in Fig. 4.2 and can be written as

$$\beta_{LH} = -\frac{\omega_0}{\omega d} \qquad\qquad \beta_{RH} = \frac{\omega}{\omega_0 d}. \tag{4.29}$$

As explained in Sect. 2.2 at low frequencies the line will behave as a PLH-TL and at high frequencies as a PRH-TL. At all other frequencies, the transmission behavior would be a combination of the RH and LH components and will result in a band gap. The limits of the band gap were given in Eqs. 2.29 and 2.30. The resonance condition of both line types for a resonator of physical length l was given in Eq. 4.1. Characteristic for an LH-line are their negative β values. Consequently, the resonances of an

LH-line resonator are attributed to negative n-values $n = -1, -2, -3, \ldots$, whereas the RH-line resonances correspond to values $n = 1, 2, 3, \ldots$ Using Eq. 4.29 and the relation that the number of cells n_C is equal to l/d, the resonance frequency for an RH-TL was derived in Eq. 4.4 and for the LH-TL case in Eq. 4.7. Furthermore, a comparison of both resonance frequencies in terms of changes in the capacitance of the unit cell was also derived and given in Eq. 4.10. As mentioned previously the number of the harmonic n and the number of used cells n_c can be chosen to be equal to profit from the use of LH-TL and therefore obtain a resonance frequency of the LH-TL resonator of nearly 10 times lower than that of the corresponding RH resonator, both with the same physical length l.

Since the sensing of a MUT is done by tracking the resonance frequency, the accuracy can be improved by having a high quality factor Q resonator. From textbooks, the value of the unloaded line resonator Q_0 is well known:

$$Q_0 = \frac{\beta}{2\alpha}. \tag{4.30}$$

It should be noted that it is the same for all harmonics. Assuming that the inductor and capacitor have quality factors of Q_L and Q_C, respectively, and employing the relations:

$$\begin{aligned}
\gamma &= \alpha + j\beta \\
\gamma_{LH} &= -\frac{j}{\omega d\sqrt{L\left(1+\frac{1}{jQ_L}\right)C\left(1+\frac{1}{jQ_C}\right)}} \\
\gamma_{RH} &= \frac{j\omega}{d}\sqrt{L\left(1+\frac{1}{jQ_L}\right)C\left(1+\frac{1}{jQ_C}\right)},
\end{aligned} \tag{4.31}$$

leads to the result for the unloaded Q factor for all harmonics and both line types:

$$Q_{0,LH,RH} = \frac{1}{\left(\frac{1}{Q_C}+\frac{1}{Q_L}\right)}. \tag{4.32}$$

In summary, the use of LH-TL cannot be motivated by the perspective of higher resonator Q_0 factors but by the possibility to strongly reduce the operation frequency. Since for adequate sensor operation, the physical dimensions of the sensor and the unit cell length must be significantly smaller than the wavelength, this approach is superior to an RH resonator. Secondly, the unit cell configuration of the LH-TL with the variable element in the serial branch can be applied much better to a sensing transmission line than the RH one where the capacitor is in the parallel branch.

4.3.2 Transmission Line Mode of Operation

The second mode of operation is the transmission line approach for monitoring the spatial distribution of the MUT. The theory on how a CRLH-TL behaves in the transmission line approach was thoroughly discussed in Sect. 2.2. Before the extraction techniques presented in Chap. 3 were used on the sensors presented throughout this work, the time domain reflectometry method (TDR) was initially used to calculate the position of an object in the vicinity of a transmission line. Therefore, a short rise time pulse is inserted in one end of the waveguide and is partially reflected when it encounters discontinuities on the line. Additional reflections occur at the beginning and the end of the line due to the discontinuities between the sensor element and the connectors. Every impedance variation along the line will result in a partial reflection of the pulse. Since the location of the discontinuity is measured by the time a pulse takes to travel to the discontinuity and back, the amplitude of the reflected pulse is determined by the time dependent reflection coefficient:

$$s_{11}(t) = \frac{Z(v_g, t) - Z_0}{Z(v_g, t) + Z_0}, \tag{4.33}$$

where Z is the input impedance at the point of reflection and Z_0 the characteristic impedance of the line. Since the speed of the pulse propagation v_g is constant along the line, the distance to the reflection point can be estimated by measuring the reflections over time. For the case of the examined transmission line, the amplitude of the reflected pulse varies differently depending on the discontinuities found in the line which are generated by mismatching impedances along the line. From Eqs. 4.28 and 4.29 the group velocity can be calculated with:

$$v_{g,LH} = \frac{\partial \omega}{\partial \beta} = \frac{\omega^2}{\omega_0} d. \tag{4.34}$$

The broadness of one pulse in relation to the length of one unit cell can be calculated with [13]:

$$n_{c,pulse} = \frac{v_{g,LH} \cdot \frac{1}{\Delta f}}{d} = 2\pi \frac{\omega_c^2}{\omega_0 \Delta\omega}, \tag{4.35}$$

where ω_c is the center frequency of the pulse with bandwidth $\Delta f = \Delta\omega/2\pi$. This value determines the capability to separate objects. The issue of time resolution and broadness of the reflected pulses is of great importance as it influences directly the achievable spatial resolution. As the electric length of the line increases, the pulse becomes broader due to dispersion effects. To obtain the location of the altered capacitance, the amount of reflected and transmitted energy over time has to be plotted. At the moment of maximum transmitted energy $t_{s_{21}}$, the center of the pulse

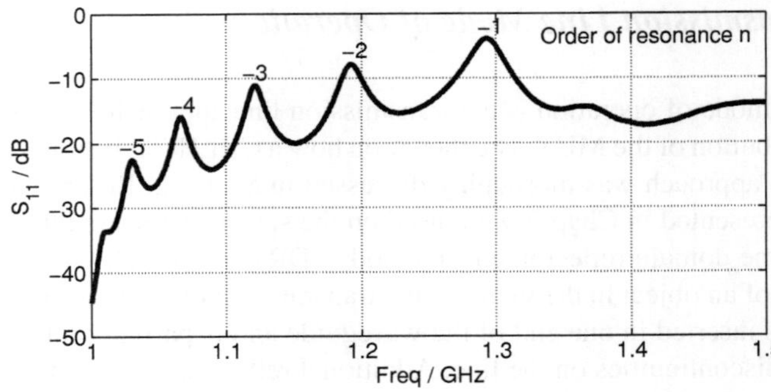

Fig. 4.6 Results from full-wave simulation of the structure depicted in Fig. 4.5c

has traveled along all n_c cells. From this we get the time the pulse needs to cross one cell:

$$t_c = \frac{t_{s21}}{n_c}. \tag{4.36}$$

The point in time of maximum reflected energy $t_{s_{11}}$ is the time needed for one reflected pulse to travel forth and back. Thus we can determine the cell, where the reflection took place:

$$n_r = \frac{t_{s11}}{2} \cdot \frac{1}{t_c} = \frac{t_{s11} n_c}{2 \cdot t_{s21}}. \tag{4.37}$$

To better understand the reason for selecting an LH-TL, a comparison between the time delays associated with an LH and an RH-TL is necessary and was given in Eq. 4.23. From this result is clear that the LH-TL has a larger delay than the RH-TL delay when the same lumped elements are considered resulting in a length reduction of the sensor element but when the line is large or the bandwidth is broad the dispersion is strong and this unfortunately reduces the resolution.

4.3.3 Sensor Design

The implementation of the planar sensor schematics shown in Fig. 4.5 into a physical layout has been conducted with the software CST Microwave Studio. Major design issues were the line impedance of 50 Ω, and a sufficient sensitivity to objects in a distance up to 4 mm. The simulated reflection of the structure for the LH band is depicted in Fig. 4.6 and the resonances are shown. It should be noted that the resonances are in ascending order since this structure behaves as a LH-TL at this frequency range. Unfortunately the RH band is not clearly visible in the reflection parameters and was suppressed from this figure. The reason for this was extensively

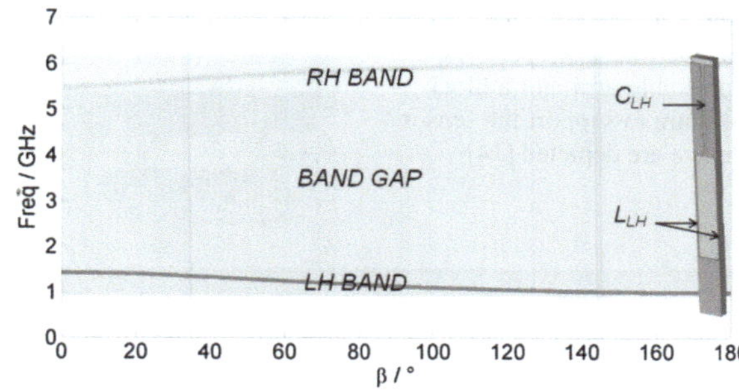

Fig. 4.7 Dispersion diagram from one unit cell of the planar microwave sensor. This diagram was extracted from the eigenmode solver of CST Microwave Studio. The LH and RH band are clearly visible

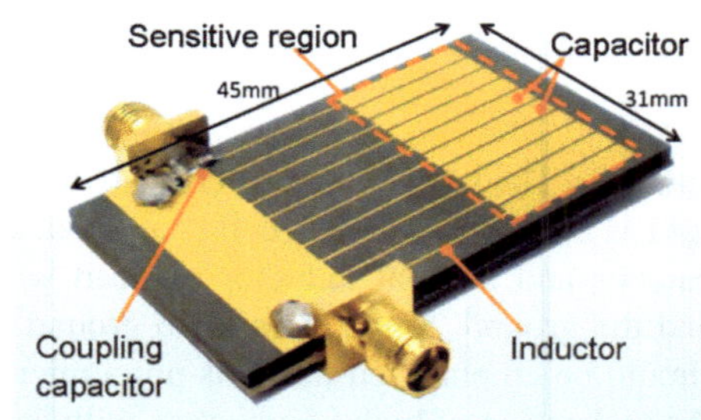

Fig. 4.8 Prototype of the planar microwave sensor. It contains 9 unit cells made of long thin line to represent the L_{LH} and patch elements to represent the C_{LH} (the dimensions in mm are highlighted) [14]

analyzed since the RH band is theoretically always present. As seen in Sect. 2.1, the dispersion diagram is the appropriate tool to observe the propagating bands. To simplified the process a single unit cell was simulated using the eigenmode solver from the software CST Microwave Studio and the results are depicted in Fig. 4.7. The dispersion diagram is not given in terms of the angular frequency ω but in terms of the frequency f to make a direct comparison with Fig. 4.6 and check if the LH band is located in the same frequency range. As seen from Fig. 4.7 the LH Band is indeed located between 1 GHz and 1.3 GHz and the RH band do exist in the structure and is located between 5.5 GHz and 6 GHz. The expected band gap is located between the two bands. Since there is proof that the RH Band exist, the next step is to analyzed why it is not visible in the simulated reflection of the structure. After considering several possibilities the most probable reason is due to the nature of the excitation of the structure that behaves as a low pass filter. Typical CRLH-TL structures are excited through the C_{LH}, in the case of the structure at hand this was not possible since the idea was to include the ground plane on the same surface and therefore the structure is excited through the L_{LH} by a CPW line connected to a coaxial SMA connector and they are most probably the cause why the RH band is not visible in the simulated reflection of the structure.

The final prototype sensor element is depicted in Fig. 4.8. It was built on Rogers RT/Duroid 5880 with a dielectric constant of $\varepsilon_r = 2.2$ and a thickness of $h = 1.57$ mm. The bigger metal patch at the top is the ground plane. As mention previously, two feeding CPW transmission lines in this ground plane soldered with two 3.5 mm SMA connectors are used to connect the sensor to a vector network

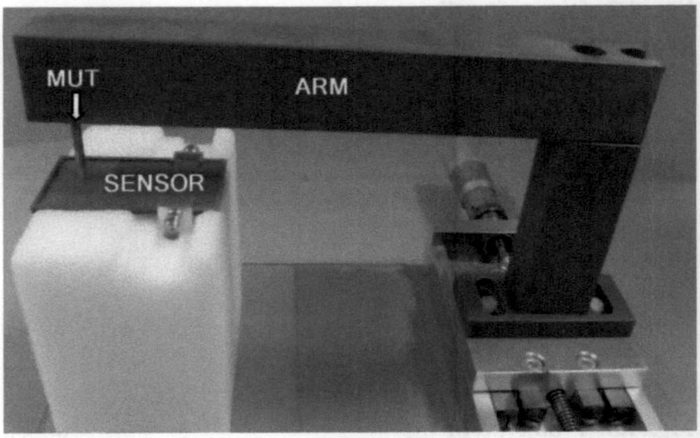

Fig. 4.9 Measurement setup for the planar sensor. The adjustable arm and a block of foam to support the sensor device are depicted [14]

analyzer. The sensor element consists of a periodical arrangement of $n_c = 9$ identical LH unit cells and each cell consists of a serial capacitor realized by two metallic patches and a shunt inductor realized with a thin long line between the patches and the ground. Since there is no ground plane on the backside of the substrate it should be emphasized that it is not a microstrip circuit. This configuration enables the realization of high Q inductors with comparably large inductive values.

Since different measurements were desired, where the MUTs are statically located on the sensor or sweept across the sensitive area, a special setup was needed. The realized test facility is shown in Fig. 4.9. It includes an adjustable plastic arm to hold the MUT and a block of foam to hold the sensor itself which slightly shifted the resonance frequencies. The movement of the arm is controlled by a micrometer screw. With this setup the sensor remains steady throughout the measurements and the MUT can be precisely located on top of the sensitive area of the sensor.

4.3.4 Measurement Results

The extraction of relative changes of the permittivity in the MUT is possible due to reference measurements where the phase change is monitored for materials with known permittivity and a linear approximation is performed. The MUT must be homogeneous and no air gap is considered. Due to the linearity of the model, the method is still applicable for other materials with different unknown values of their permittivity. Using the test facility several measurements where done always detecting the phase change according to the relative permittivity of the MUT. First of all, several dielectric plates with different known relative permittivities were placed on top of the sensor and the absolute phase was measured. This procedure was done for four consecutive modes in the LH band as shown in [14]. The biggest overall change in the phase appears for the first mode. Nevertheless, between $\varepsilon_r = 1$ and $\varepsilon_r = 2.2$ the curves for all four modes are quite steep, indicating a higher sensitivity of the sensor for small changes in ε_r independent of the chosen mode. For higher changes

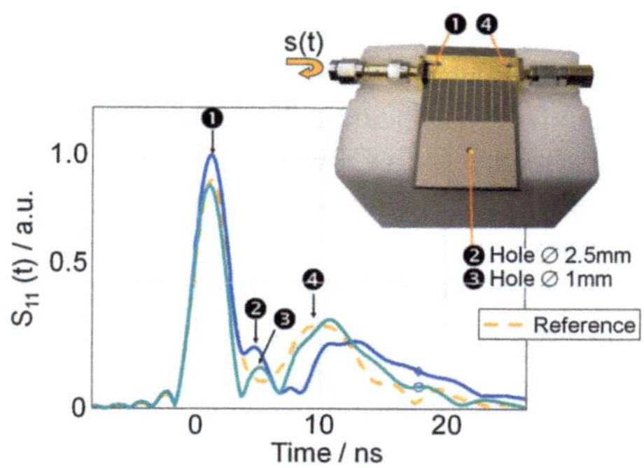

Fig. 4.10 Broadband time domain measurements with dielectric plate. A dielectric plate is located on *top* of the sensor with two holes of different diameters in the *center*. The position and size of the holes can be extracted from the graph [12]

between $\varepsilon_r = 2.2$ and $\varepsilon_r = 10$, the first mode shows a bigger phase difference and all the other modes have similar less steep behavior. From this analysis, it can be concluded that the first resonance mode should be used in the extraction of the dielectric properties of the MUT.

Other measurements where carried out by creating a perturber or disturbance in the MUT in order to corroborate if the sensor is able to detect them. For this purpose, holes of different sizes were created on the dielectric plates. An exemplary result is shown in Fig. 4.10 where two different measurements, each with a dielectric plates that contain a hole of certain size, are overlapped for an easy comparison. The dielectric plate has $\varepsilon_r = 10.8$. Here the time domain S_{11} parameters are plotted, first the reference with no MUT on top of the sensor and two peaks are clearly detected that correspond to the reflection caused by the discontinuities between the sensor element and the two connectors respectively. When the dielectric plate is placed on top of the sensor a new reflection appears at an specific time depending on the size of the hole. The peak marked as number 2 corresponds to the measurement where the dielectric plate had a larger hole of 2.5 mm located in the middle and the peak marked as 3 corresponds to the plate with a smaller hole of 1 mm also locate in the middle of the plate. As expected, the different sizes of holes create a reflection on the line at different times enabling its identification.

The last measurements included medicine tablets placed at different positions on top of the sensor, where a similar behavior as the one experienced with the dielectric plates can be clearly identified for two different positions of the medicine pills. The results are shown in [12]. From the location of the peak the position of the medicine pill can be approximated. From the theoretical point of view, the group velocity can be calculated from the reference measurement and has a value of 5.6×10^6 m/s. Since the number of unit cells and its size is known, the delay time of the reflection pulse can be calculated as follows:

$$t = 2\left(\frac{d}{v_g}\right)n = 4.9\,\text{ns}. \tag{4.38}$$

The value from Eq. (4.38) is comparable with the measured value that corresponds to a time delay of 5.1 ns. The calculated and measured values are in good agreement, the difference is due to the fact that the calculations are performed for a pure LH-TL and the actual measurements are for a CRLH-TL.

In order to extract the position of an object placed on top of the planar sensor, the broadband time-domain analysis was used. From the measurements, it is clear that when the pulse reaches the medicine tablet or the hole in the plate, it sends a reflection pulse that can be seen in the time dependant reflection coefficient and it is relative to the objects position. From the delay and amplitude of this pulse the position of the MUT can be derived. The reason for using a well known time domain technique is first of all its simplicity. Also it was necessary to test if this method was sufficient for extracting the position of an object in the vicinity of the sensor with one unit cell resolution. Although the method gives promising results as previously seen, it unfortunately does not provide the desired resolution. When two separate MUTs are in neighboring unit cells the TDR extract them as one MUT. On the other hand, TDR is not able to directly extract the relative changes in the dielectric constant and the use of the sensor in resonant mode with phase monitoring is necessary for this purpose.

In summary, the presented sensor concept can detect relative changes in the dielectric constant of the MUT and furthermore its position thanks to the use of a periodic structure. The broadband time domain measurement can help extract this information but in a limited way since it does not provide a one unit cell resolution or a spatially resolved profile with the same sensor device. It requires alterations on the sensor to operate in two modes depending on what information is desired, position or relative changes in the dielectric properties of the MUT. With more robust extraction methods like the MRPM and MCEM presented in Chap. 3, it is possible to obtain the relative changes in the dielectric constant and the position using the same structure with no special coupling and a one unit cell resolution can be achieved. In the next section a redesign of this sensor concept will be presented and a new design based on a coplanar waveguide geometry for the intended application of paper monitoring will be also addressed. The advance extraction methods will be used and even measurements with high moisture content will be presented.

4.4 Detection of Defects and Moisture in Paper

As mentioned previously, the chosen application to test the principle of a planar metamaterial sensor in an industrial environment is for the paper industry. The goal is to monitor contact-less the quality of raw paper and its moisture content. Two different designs were considered. The first geometry was presented in Sect. 4.3.3. The structure has an unconventional geometry with the ground plane on the same level as the sensing part but it is not a microstrip or CPW design. The second sensor to be designed and realized is a traditional CPW TL made of a periodic arrangement of unit cells. Measurements with both sensor prototypes were executed to establish

Table 4.1 Parameters for the design of a CRLH-TL sensor for paper quality monitoring in an industrial environment

Parameter	Desired value
Center operation frequency	2.5 GHz
Sensor length	30 cm
Number of unit cells	$3-5$
Sensitivity $(\Delta\varepsilon/\varepsilon)_{min}$	5 %

the concept. A material identification is conducted and distance measurements up to 2.5 cm with four different kinds of paper are performed.

Two different methods for extracting the capacitive profiles out of the sensor structure are used from Chap. 3. The first method is the MRPM where the information is taken from the frequency shifts of the line's input impedance. With this method the determined frequency shifts could be used to reconstruct the capacitive profile along the transmission line. The second method is the MCEM, which is a combination of the classical network synthesis Cauer canonical forms. The method construct a impedance transfer function and then performs a continued fraction expansion for extracting all lumped elements of the equivalent circuit for each CRLH unit cell.

4.4.1 Sensors Designs

The design of the two CRLH-TL is carried out with the software package CST Microwave Studio. The major design objectives for both prototypes are presented in Table 4.1. The reason for the selected value of the design parameters will be discussed. The operation frequency and the sensor length was selected to obtained large structures that could test $A5$ paper. Also, for the application at hand, a larger sensor is more suited since it could be mounted on a conveyor belt during the paper production cycle. Most conveyor belts can have a width of up to several meters and since the sensors have to be located closed to the MUT, their size should be in the range of the size of the MUT. Nevertheless, as mentioned previously, one big advantage of the sensor concept presented in this work is the high flexibility in terms of size, operation frequency and geometry that it offers. The next parameter, amount of unit cells, was selected to be not very large to minimize problems with the extraction techniques in this first proof of concept. Finally, the desired sensitivity is quite high to be able to detect slight changes in the properties of the paper and thinking ahead that moist paper measurement may need a higher sensitivity due to the inherent lossy properties of that MUT and therefore the difficulties that arise for sensing in the microwave range where water molecules have a high lossy behavior.

4.4.1.1 Sensor Design 1: Planar Sensor Prototype

The first sensor consists of a planar metamaterial transmission line. In Fig. 4.11 the fabricated prototype is shown. It is implemented on the substrate Isola FR408, which

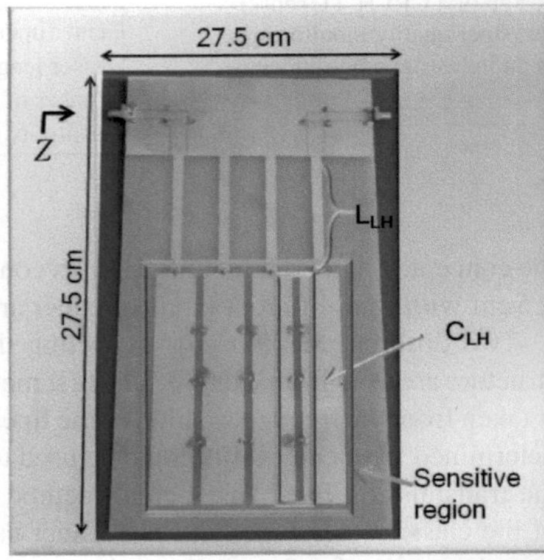

Fig. 4.11 Sensor prototype 1 for the paper application (the dimensions are highlighted in cm)

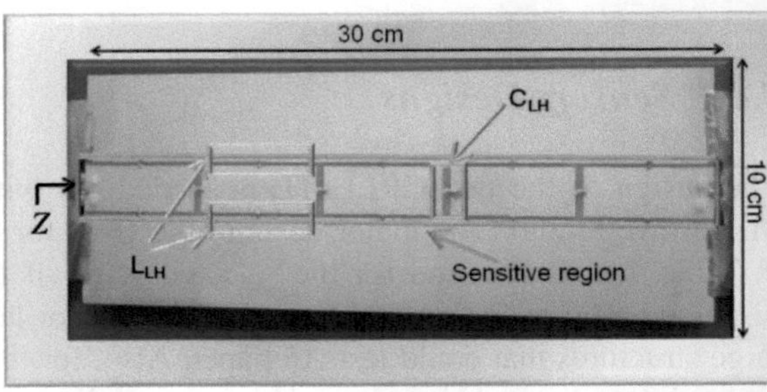

Fig. 4.12 Sensor prototype 2 for the paper application based on a CPW structure (the dimensions are highlighted in cm)

has a dielectric constant $\varepsilon_r = 3.75$ and a height of 1.75 mm. The sensor element consists of a periodical arrangement of four identical unit cells. Each cell is build up of a serial capacitor realized by two metallic patches, a shunt inductor realized with a long thin line between the patches and the ground, the RH components are obviously present as a parasitic effect. On the top of the sensor is a rectangular patch, which represents the ground plane. In this ground plane two feeding CPW transmission lines are used for connecting the sensor to a network analyzer. They are soldered to two 3.5 mm SMA connectors. Additional lumped elements are used to better control the behavior of the sensor as a CRLH-TL and its operating frequency. The distance between the patches that represent the C_{LH} is quite large to make the field protrude farther into the MUT and therefore the lumped elements are necessary to bring the frequency down to the desired value.

4.4.1.2 Sensor Design 2: CPW Sensor Prototype

The layout of the second sensor is illustrated in Fig. 4.12. The sensor consists of five unit cells serially arranged and connected with each other forming a CRLH-TL implemented in a coplanar waveguide. For adjusting the operation frequency and also the sensitivity of the sensor, SMD-inductors and capacitors are inserted into the design as was done with the planar sensor 1. The capacitors C_{LH} are always positioned between the individual cells. On both sides of a sensor cell the inductors L_{LH} are placed between the cell patch and the ground. It is implemented on the substrate Isola FR408, which has a dielectric constant $\varepsilon_r = 3.75$ and a height of 1.75 mm.

4.4.2 Measurement Results

In the following section, the measurements of the realized sensors are analyzed and compared to the corresponding simulation results with different MUTs. The MUT have to be homogenous and no interferences like air gaps are considered. In the first measurement, the frequency domain reflection coefficient S_{11} is determined. From the S-parameters the real and imaginary part of the line input impedance Z can be extracted as needed for the extraction methods. Since measurements at different distances are required, an appropriate setup is constructed as seen in Fig. 4.13. It includes a stationary plate and a moveable plate made of PVC with four rods, one on each corner. For measurements where the MUT can be placed directly on top of the sensor, the sensor can be located as seen in this figure. For measurements where the distance between the sensor and the paper needs to be adjusted, the sensor can be secured on the opposite side of the moveable plate, so that the sensitive region is facing downward as seen in Fig. 4.14. The moveable plate can be raised up or down manually and the MUT is placed on the lower plate. Additionally, micrometer gauges are placed in each corner of the setup to control the distance between the plates during upward or downward movement.

The first step is to characterize the different MUTs that will be used. Four different kinds of paper are used as MUT as shown in Fig. 4.15. The papers vary in configuration, shape and thickness. They all possess a permittivity close to $\varepsilon_r = 2.31$.

In the tests made for both structures the same procedure was followed. First, a reference measurement of the reflection coefficient S_{11} is done without any MUT i.e empty measurements of the structures. Second, calibration measurements were done since they are needed for making the extraction with the MRPM. They are done by covering the sensitive region with the MUT perturbed by holes on each unit cell respectively. The calibration for the planar sensor prototype is shown in Fig. 4.16 and a similar procedure for the CPW sensor was performed. Then different profiles were done by making holes of different sizes and shapes on the MUT. Finally, moisture measurements were performed with moist paper also with holes on different positions to create defects on the paper.

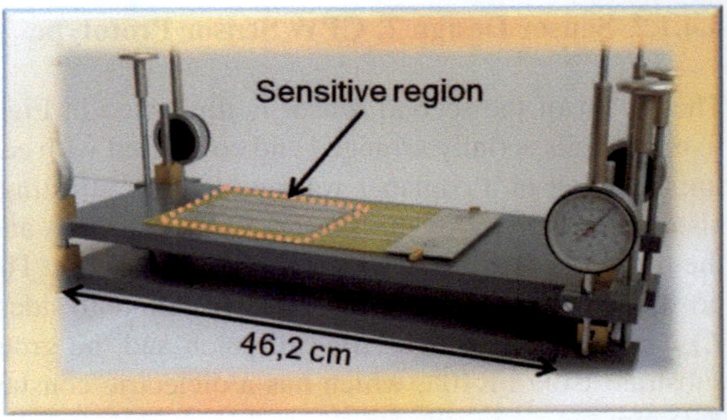

Fig. 4.13 Front view of the measurement setup for the paper sensors. It includes a stationary plate and a moveable plate with four rods. The distance between the two plates can be adjusted manually and micrometer gauges help to control that all rods are equally positioned

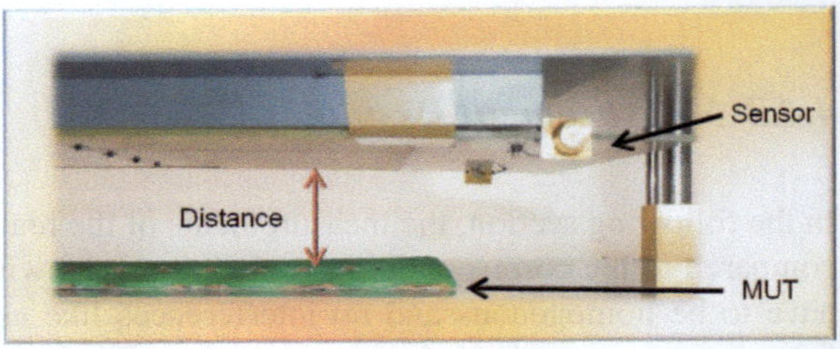

Fig. 4.14 Measurement setup for the paper sensors when used for distance measurements. Four rods are used to adjust the *height* between the sensors and the MUT

Fig. 4.15 Paper samples used as MUT. From *left* to *right* pink board paper, laminated *orange* board paper, corrugated *green* board paper and grey board paper

4.4.2.1 Measurement Results for Sensor 1: Planar Sensor Prototype

The measurement results for the planar sensor prototype will be discussed in this section. In a first step the reflection coefficient S_{11} of the simulation and measurements without MUT are plotted. A very good agreement could be achieved between simulation and measurement as depicted in Fig. 4.17. Three peaks indicating the resonances can clearly be detected. Additionally, for extracting the resonances, necessary in the used of the extraction methods, the real part impedance $\mathrm{Re}(Z_L)$ is seen in Fig. 4.18 and the imaginary part impedance $\mathrm{Im}(Z_L)$ is seen in Fig. 4.19. In theory

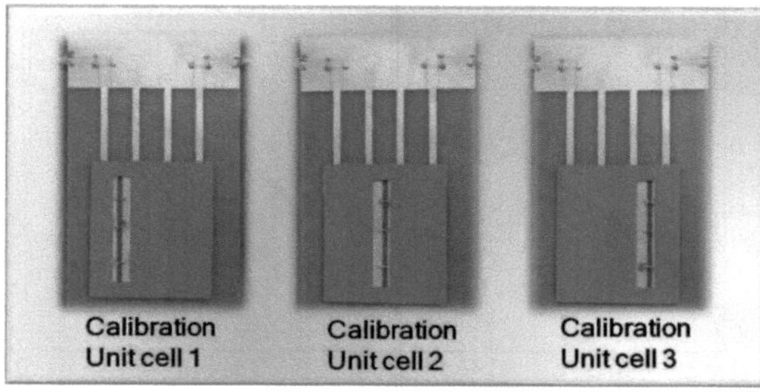

Fig. 4.16 Reference measurements with the planar sensor prototype

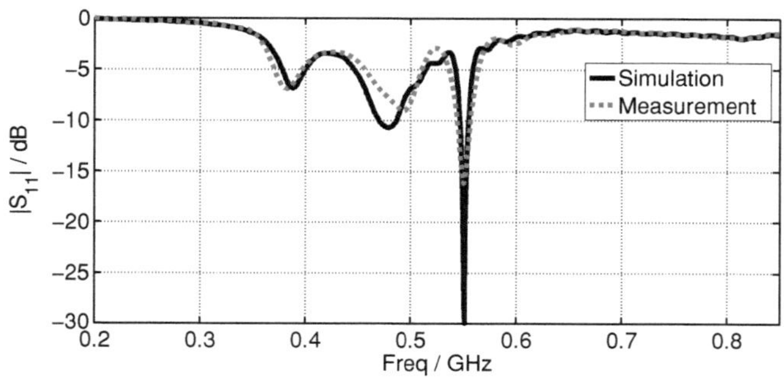

Fig. 4.17 Comparison between simulation and measurements for the planar sensor prototype for monitoring quality of paper

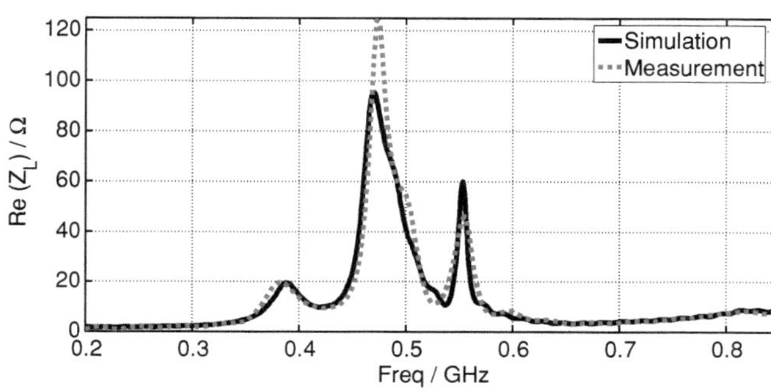

Fig. 4.18 Real part of the impedance of the planar sensor prototype for monitoring quality of paper

the resonances can all be extracted from the $\text{Im}(Z_L)$ by searching the location where the function crosses the zero line, but in practice this is not always possible since the $\text{Im}(Z_L)$ tends to drift away from the zero line mainly due to the losses within the MUT or the system itself. Therefore the use of the $\text{Re}(Z_L)$ is necessary and the poles and zeroes of the Z_{in} can be located simply by searching the maximums in the $\text{Re}(Z_L)$ and $\text{Re}(Y_L)$ respectively.

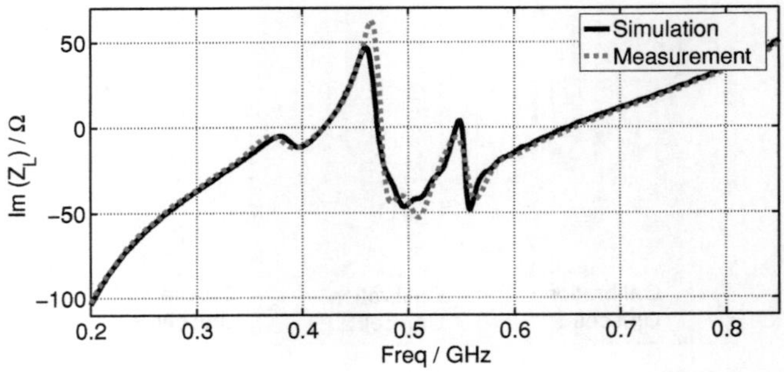

Fig. 4.19 Imaginary part of the impedance of the planar sensor prototype for monitoring quality of paper

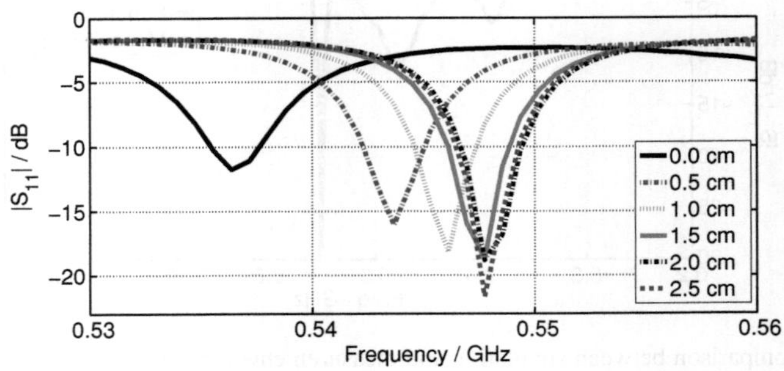

Fig. 4.20 Exploit view of the first resonance frequency where distance measurements with the planar sensor prototype of *grey* board paper are depicted

The results for the distance measurements are illustrated in Fig. 4.20. A closer view of the higher resonance frequency can be seen in this figure to better appreciate the changes produce by the loading of the different paper samples. As reference measurement the MUT is directly placed on top of the sensor. With increasing distance to the MUT the capacitive loading $\Delta\varepsilon_r$ decreases, which results in a shift $\Delta\omega$ to a higher frequency.

For a better comparison of the changes in frequency for the individual distances, a figure of merit is defined as follows:

$$FoM = \frac{\Delta f_x}{f_x}, \text{ in } \%, \tag{4.39}$$

where f_x is the current resonant frequency and Δf_x describes the difference between the current resonance frequency and the resonance frequency obtained by the next measurement performed with a higher distance between the sensor and the MUT. The desired limits of the figure of merit is between 0.1 and 10.0 %. Within these limitations, the sensitivity is sufficient for indicating objects. The results are present in Fig. 4.21 and from the reference point at 0.0 cm to the maximum distance step

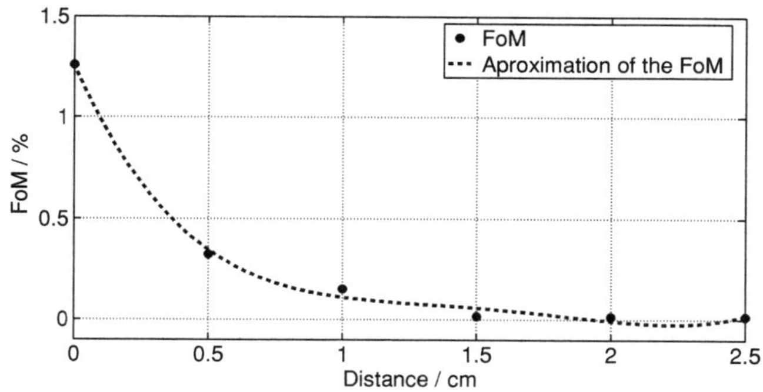

Fig. 4.21 Analysis of the FoM for the planar sensor prototype 1 during measurements at different distances of the MUT

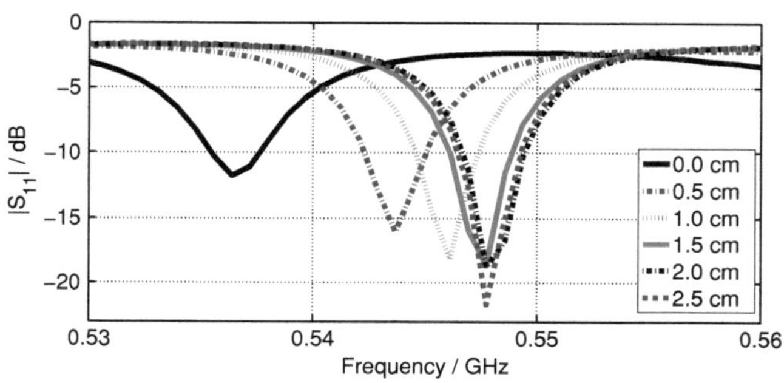

Fig. 4.22 Exploited view of the first resonance frequency where distance measurements with the CPW sensor prototype of *grey* board paper are depicted

of 2.5 cm the sensitivity is falling steadily. In the last distance step up to 2.5 cm a frequency shift of 1.0 MHz can be recognized for the structure. The lower acceptable limit is then reached in a distance of 2.5 cm. It can therefore be concluded that with an increasing distance the sensitivity of the sensor structure decreases with a nearly exponential behavior.

4.4.2.2 Measurement Results for Sensor 2: CPW Sensor Prototype

The distance measurements of the second sensor design are presented in Fig. 4.22. As well as for the previous design, a closer view of the higher resonance frequency is depicted and the reference measurement is done by placing the MUT directly on top of the sensor. With increasing distance to the MUT one can also recognize a frequency shift $\Delta\omega$ to a higher frequency, because of the loading effect, which is described by the decrease in the overall ε_{eff}. By looking at the figure of merit for the sensitivity presented in Fig. 4.23 the following conclusions can be drawn.

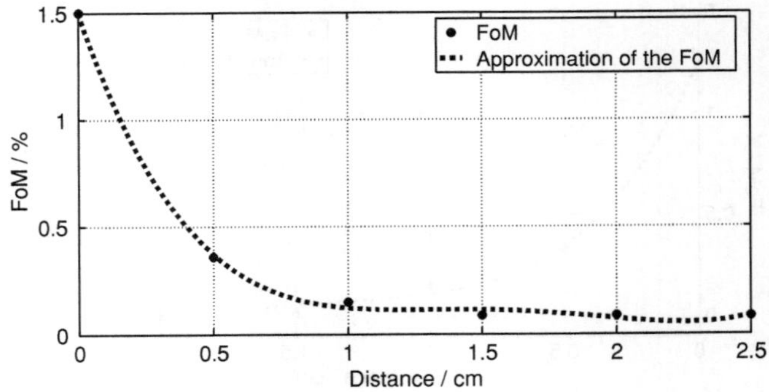

Fig. 4.23 Analysis of the FoM for the CPW sensor prototype during measurements at different distances of the MUT

The sensitivity decreases as the distance between the sensor and the MUT increases. Compared to planar sensor structure 1 the change in frequency is slightly higher. A comparatively small frequency shift of 0.03 % (1 MHz) can be observed. A more detailed comparison of the two sensors in terms of sensitivity will be addressed.

4.4.2.3 Comparison Between the Two Sensor Designs for Paper Quality Monitoring

To compare the presented structures their sensitivity to distance measurements is analyzed. The sensitivity is here defined as the ability to detect an object and localize its position as accurately as possible. As a result of the measurements, it could be established that additional objects, which were placed outside of the sensitive region, do not influence the actual application. This applies for both structures.

Overall, the sensitivity of the CPW sensor 2 is always above the sensitivity of the planar sensor 1 but for both structures it decreases in a similar way and a frequency shift of 1 MHz can be recognized for both structures. As mentioned before, the lower limit, where the identification of objects is still possible, is defined as 0.1 %. With both sensor designs this limit is reached in a distance of 2.5 cm. This leads also to the conclusion that for distances up to 1.0 cm CPW sensor 2 is more qualified to detect any irregularities.

In summary it can be said that both sensor designs fulfill the initial requirements and can be used for monitoring MUTs. Dependent on the application and the technical requirements that the sensor has to address, one or the other design can be chosen. For dynamic processes, where the conditions change constantly, CPW sensor design 2 would be more effective for an observation of objects, because it has a higher sensitivity related to the distance. For the differentiation and characterization of individual MUT, planar sensor 1 would be more appropriate. The disadvantages of the sensors lie on the fact that SMD elements are used, which are soldered onto the structure, and have a strong influence on the transmission line properties. Therefore, this variation could have an influence on the overall dielectric change. Additionally, potential deviations in the measurements, coming from the complete system where the sensor is located, cannot always be avoided. Finally, during the measurement an air gap is always present between the sensor and the MUT, resulting in errors in the detection. Nevertheless the sensors performance is satisfactory and the proof of concept was established.

4.4.3 Profile Extraction

4.4.3.1 Multi-Resonant Perturbation Method

With the MRPM the capacitive profile of the two sensor designs can be reconstructed as seen in Sect. 3.1. In theory the capacitive profile ΔC_{LH} can be calculated from the relative resonance frequency shifts $\Delta\omega$ multiplied by the inverse of the calibration matrix A^{-1}. But as it will be shown below, in practical applications the condition of the matrix is sometimes not appropriate. The precise knowledge of the A-matrix is essential for a good sensor performance. The perturbers as already mentioned, are holes placed at different positions in the MUT. Consecutively the calibration measurements are performed and all capacitances $C_{LH,n}$ are set to the value one in the initial adjustment. If one capacitor $C_{LH,n}$ is disturbed, a capacitive change $\Delta C_{LH,n}$ occurs while all other capacitances $C_{LH,m}$ have the value one $(m \neq n)$.

For determining the capacitive profile several measurements with different perturbers were executed. In Fig. 4.24 an example of the extracted profile for planar sensor design 1 is illustrated. Here, the disturber is placed at the third unit cell and the profile looks reasonable but the value for the unit cell 2 is too high, nevertheless, a good agreement with the desired profile is obtained. The discrepancies from the theoretical conditions have several possible reasons, for example they could be the result of the simplified assumptions made in the MRPT where a linearization of the problem is used or a problem in the extraction of the correct resonance frequencies. During the measurements non-linear effects coming from the measurement setup can gradually decrease the precision of the detection. In Fig. 4.24 the change of the dielectric constant of each unit cell $\Delta\varepsilon_n$ is normalized to the calibrated dielectric constant change $\Delta\varepsilon_{cal}$ and is plotted for each unit cell. The graphs illustrate the capability to have a one cell resolution.

A profile extracted from the CPW sensor 2 is shown in Fig. 4.25. In this example the disturber is placed at unit cell 3. As it can be seen, for this sensor design, the extracted profile is satisfactory. The same small errors seen with the other sensor design are still there but the profile can be identified.

Overall, all pertubers can be localized correctly with both sensor designs, because they have a strong influence to the corresponding resonance frequency. As mentioned before, there are still some deviations from the ideal profiles. The extraction accuracy is limited by external influences that can appear during the measurement.

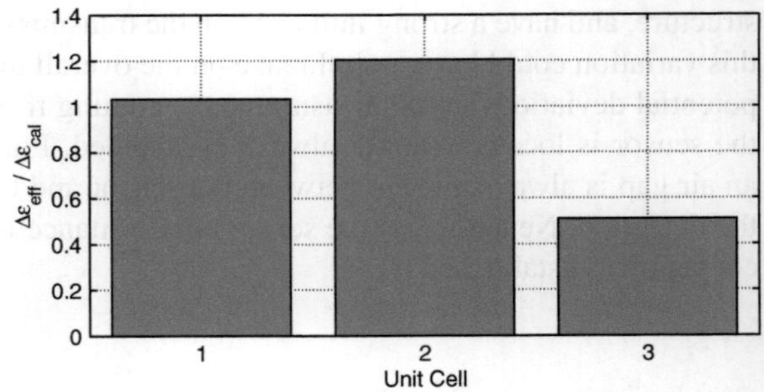

Fig. 4.24 Extracted profile with the prototype planar sensor 1 using the MRPT. The perturber is located in the unit cell 3

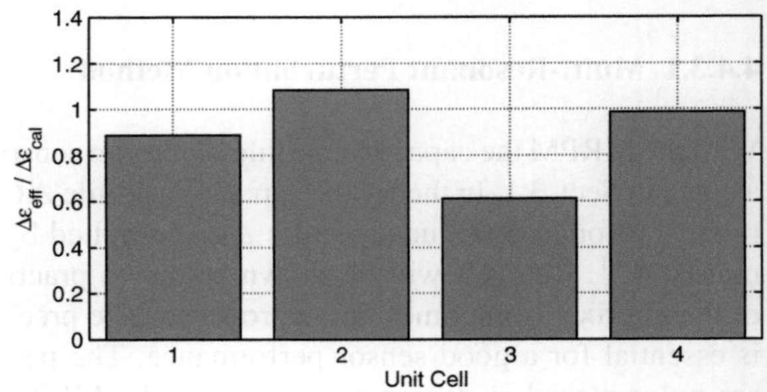

Fig. 4.25 Extracted profile with the prototype CPW sensor 2 using the MRPT. The perturber is located in the unit cell 3

Atmospheric moisture or small air gaps between the MUT and the sensor can influence the results significantly. However these graphs demonstrate the operation principle and proof of concept using the MRPM.

4.4.3.2 Mixed Cauer Expansion Method

For extracting the equivalent circuit of the complete sensor, the MCEM is used only for the CPW sensor structure 2. The reason is that contrary to the planar sensor structure 1, the S-parameters of this structure show an observable LH as well as a RH band. In Table 4.2 all the extracted values for both LH and RH elements are compiled for the sensor structure loaded completely with paper and no holes or perturbers are included.

The same extraction procedure was done with the sensor loaded with different profiles. In Fig. 4.26 the profile of the sensor completely covered with paper as seen in Table 4.2 is plotted with a perturber hole made on the second unit cell. The values of the undisturbed unit cell have a large error, they should all have a value of approximately 1. Nevertheless, the perturbed unit cell still has the smallest value. The reason for the large errors was already explained in Chap. 3. The MCEM has severe accuracy problems when the structure has an amount of unit cells larger than two and when real measurements are considered. The problem seems to be the algorithm itself where continued divisions of the polynomials are performed. In real measurements

Table 4.2 Extracted values of the sensor CPW prototype equivalent circuit using the MCEM.

Unit cell	L_R(nH)	C_L(pF)	C_R(pF)	L_L(nH)
1	2.18	4.73	6.81	1.85
2	2.12	3.88	6.47	1.73
3	2.03	3.34	6.50	1.65
4	1.99	3.28	6.29	1.55
5	1.97	3.26	5.95	1.43

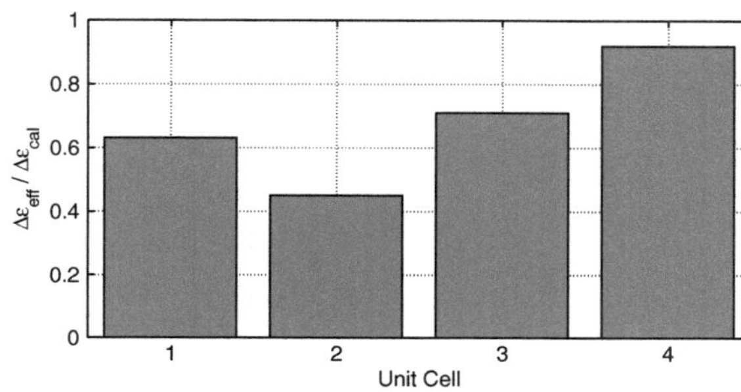

Fig. 4.26 Extracted profile with the prototype CPW sensor 2 using the MCEM. The perturber is located in the unit cell 2

this is worst since the imaginary part of the input impedance is shifted from its center due to losses from the MUT or the system itself. In future work, stronger signal processing mechanisms should be consider such as optimization techniques to improve this method. Since the MRPM shows a better performance under real conditions, it will be used for extracting the profiles for the wet paper measurements that will be presented in the next section.

4.4.4 Measurements with Moist Paper

A very interesting property to monitor in the paper industry is the moisture content and as result the possible damage that can occur to the paper. To test if the sensors were able to operate adequate when coming into contact with moist paper, additional measurements were carried out. The paper used in this experiments, for both calibration and profiles, was completely moist using deionized water. In the profiles, holes were made in different locations of the paper as done with the dry measurements. In Fig. 4.27 is depicted three different profiles extracted with the planar sensor 1 using the MRPM. In Fig. 4.27a the paper has no holes but simply the sensor is fully covered. In Fig. 4.27b the paper covers the sensor completely but it has a perturber hole in unit cell 1. Finally in Fig. 4.27c the perturber hole is located in unit cell 3. All the profiles have slight errors in the absolute value but the position of the perturbers can be clearly located.

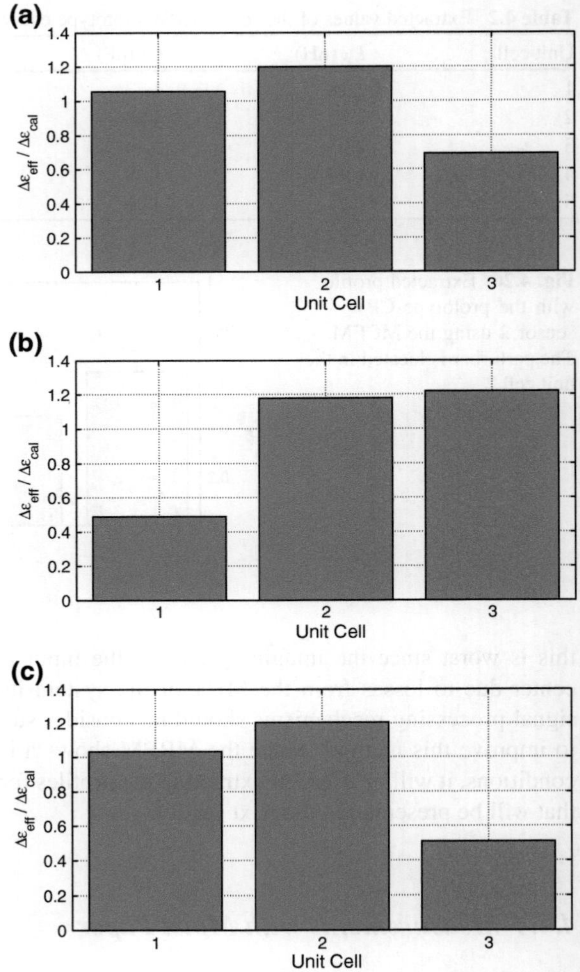

Fig. 4.27 Extracted profile of wet paper samples with the prototype planar sensor 1 using the MRPT. **a** The paper has no perturbers, it covers the complete sensor. **b** The perturber is located in the unit cell 1. **c** The perturber is located in the unit cell 3

In Fig. 4.28 are depicted the results of measurements with wet paper with the CPW sensor 2 prototype. Again the MRPM was employed since it was proven to be the more robust extraction method. In Fig. 4.28a a perturber hole is located in the unit cell 1 and in Fig. 4.28b is located in unit cell 4. The results are good but some errors are still observed. For example, in the second profile a negative value was obtained for the unit cell 4 most probably because the calibration A matrix contained slight errors. Another possible reason is that one of the poles or zeroes was not appropriately extracted because it was not clearly visible or the algorithm could simply not find it due to a wrong selection in its search parameters.

In summary moist paper measurement were successfully carried out and several profiles were extracted using both designed sensors and the MRPM. A one unit

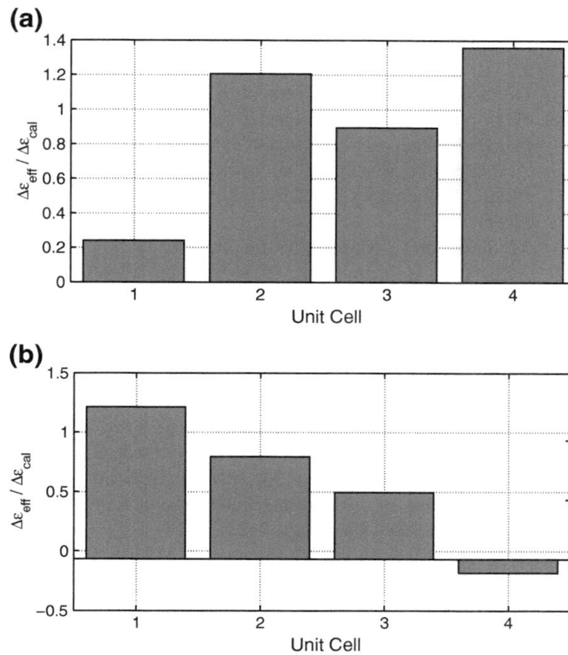

Fig. 4.28 Extracted profile of wet paper samples with the CPW sensor 2 using the MRPT. **a** The perturber is located in the unit cell 1. **c** The perturber is located in the unit cell 4

cell resolution was achieved for both sensors and the proof of concept was thoroughly made. The limit in terms of the size of the perturber that can be detected is directly proportional to the impact this perturber has on the resonance frequency. As long as it produces a change of approximately 1 % it can be detected. Another possible application of the sensor concept presented in this chapter is for moving environments, and therefore, the sensors can be mounted on conveyor belts that are commonly found in industrial scenarios. Some tests where carried out in this direction, where the sensor presented in Fig. 4.8 was mounted on a small conveyor belt and dynamic measurements were made. There it was proven that sinusoidal phase variations in the sensor-output signal for objects moving in the vicinity of the sensor are present and can be used with algorithms such as the spatial filter velocimetry to extract the velocity of objects on the belt. The findings where reported in [14] but further investigations on this topic where not pursed since it is out of the scope of this thesis. Additional work can be done in this direction to combine all the features these sensors can provide such as MUT characterization, i.e. detection of perturbers and position of the MUT and additionally velocity information. The design system presented, both sensors and extraction methods, is suitable for any industrial application where the monitoring of a dielectric change is of interest. In the following chapters the same principles will be applied but in completely different applications, the biomedical analysis and treatment, to really prove the wide range of possibilities where these sensors could operate.

References

1. D. Early, On-Line Measurement of Pulverized Coal, www.airmonitor.com, April 2007
2. H. Conrads, Method and Device for a Contact-free Measurement of the Mass Flow Rate in a two Phase Pneumatic Transport using Microwaves, EP. Patent 0 717 269 A2, November 1995
3. A. Penirschke, R. Jakoby, Microwave mass flow detector for particulate solids based on spatial filtering velocimetry. IEEE Transactions on Microwave Theory and Techniques **56**, 3193–3199 (2008)
4. M. Schüßler, C. Mandel, M. Puentes, R. Jakoby, *Metamaterial Inspired Microwave Sensors. Microwave Magazine*, vol. 13, Issue 2, pp. 57–68, March-April 2012
5. M. Puentes, A. Penirschke, M. Schüßler, R. Jakoby, Microwave Mass Flow Meter for Industrial Applications Based on Left-handed Transmission Line Resonator. In GeMiC. German Microwave Conference (Munich, Germany, 2009)
6. M. Puentes, M. Maasch, M. Schüßler, R. Jakoby, Frequency multiplexed 2-dimensional sensor array based on split-ring resonators for organic tissue analysis. IEEE Transactions on Microwave Theory and Techniques **60**(6), 1720–1727 (2012)
7. M. Schüßler, M. Puentes, D. Dubuc, K. Grenier, R. Jakoby, Simultaneous Dielectric Monitoring of Micro fluidic Channels at Microwaves Utilizing a Metamaterial Transmission Line Structure. In Proceeding of 34th Annual International Conference of the Engineering in Medicine and Biology Society (San Diego, USA, 2012)
8. C. Damm, M. Schüßler, M. Puentes, H. Maune, M. Maasch, R. Jakoby, Artificial Transmission Lines for High Sensitive Microwave Sensors. In Proceeding of IEEE International Sensors Conference (Christchurch, New Zealand, 2009)
9. C. Mandel, H. Maune, M. Maasch, M. Sasegar, M. Schüßler, R. Jakoby, Passive Wireless Temperature Sensing with BST-based Chipless Transponder, GeMiC. German Microwave Conference (Darmstadt, Germany, 2011), pp. 1101–1104
10. C. Mandel, M. Schüßler, R. Jakoby, A Wireless Passive Strain Sensor, IEEE Sensors Conference (Limerick, Ireland, 2011), pp. 207–210
11. C. Mandel, M. Schüßler, M. Maasch, R. Jakoby, *A Novel Passive Phase Modulator based on LH Delay Lines for Chipless Microwave RFID Applications*, IEEE MTT-S Int, Microwave Workshop on Wireless Sensing, (Cavtat, Croatia, 2009)
12. M. Puentes, B. Stelling, M. Schüßler, A. Penirschke, C. Damm, R. Jakoby, Dual Mode Sensor for Belt Conveyor Systems Based on Planar Metamaterials. In Proceeding of IEEE International Sensors Conference (Christchurch, New Zealand, 2009)
13. M. Schüßler, C. Damm, R. Jakoby, *"Periodically LC Loaded Lines for RFID Backscatter Applications", in Metamaterials Conference* (Rome, Italy, 2007)
14. M. Puentes, B. Stelling, M. Schüßler, A. Penirschke, R. Jakoby, Planar Sensor for Permittivity and Velocity Detection Based on Metamaterial Transmission Line Resonator. In Proceeding of European Microwave Conference (Rome, Italy, 2009)

Chapter 5
Sensing of Biological Cells

The analysis of biological cells, in order to extract important properties both from the individual cells and from pools, is usually an extensive laborious process. Most of the time, it is done by looking at the sample under the microscope and using markers or fluorescent agents that might alter the biological activity of the MUT. When a large amount of cells needs to be analyzed, the process can take an extremely long time. The aim in this chapter is to apply the basic sensor concept investigated in Sect. 4.1 into a microwave sensor array system by combining planar periodic metamaterial structures with microfluidic channels technology. Metamaterial structures were chosen due to their interesting properties for this application such as good coupling, high sensitivity and easy miniaturization. The sensor array is composed of several metamaterial unit cells, each having its own microfluidic channel, creating together the sensor elements. Therefore, measurements of several samples can be performed with this sensor array simultaneously. Thus, the analysis of the biological cells with this sensor array, e.g. in order to extract properties for diagnostic purposes such as concentration, protein function, the stage of the developing cycle, surface modification or drug delivery can be performed almost automated by measuring the dielectric properties, and hence, reduces significantly the process time. For the extraction of the dielectric profile, the multi resonant perturbation method will be used to monitor the changes of the dielectric behavior of the samples individually.

5.1 State of the Art in the Analysis of Biological Cells

Intensive research is ongoing in order to propose innovative miniature biodevices with integrated microfluidics for a wide variety of applications such as diagnostics, therapeutics, and tissue engineering. Most important they should be integrable within a lab-on-chip. These devices are used to detect cells, microorganisms, viruses, proteins, DNA and related nucleic acids, and small molecules of biochemical importance and interest [1]. The advantages of these devices lie in several factors. First,

M. Puentes Vargas, *Planar Metamaterial Based Microwave Sensor Arrays for Biomedical Analysis and Treatment*, Springer Theses,
DOI: 10.1007/978-3-319-06041-5_5,

they can be miniaturized to an appropriate size that enables the interaction with the target MUT hence providing a high sensitivity analysis. Second, they do not need reagent particles that will create a chemical reaction to permit the analysis, therefore, reducing the cost and making the procedure non-invasive to the MUT. Finally, small volumes of the MUT are enough for the diagnosis and furthermore the "Biochips" are portable.

The solutions for electromagnetic based sensors cover all the frequency spectrum. An example for low frequencies below MHz is presented in [2], where a hybrid microfluidic/IC capacitive sensor is presented for a lab-on-chip application with CMOS technology. For RF or microwaves several examples can be considered. In [3] a coplanar waveguide device integrated with a microfluidic channel performs measurements from 40 Hz to 26.5 GHz with hemoglobin solution and live *E. coli* cells. In [4, 5], another RF biosensor with a microfluidic network is presented. It makes RF/microwave detection of HUVEC human cells in different concentrations. As for the Terahertz range an example is presented in [6] where differential time-domain spectroscopy (THz-DTDS) is used to measure minute changes of bovine lung microvessel endothelial cells in response to vascular endothelial growth factor. The results were taken from the alterations in the attenuation of the wave and therefore the changes on the THz dielectric properties of the studied cells. In [7] a THz probe integrated inside a microfluidic BioMEMS is also presented. In optics there is also work being done for electrical and mechanical devices as presented in [8]. Here it is presented the recent advances in the electric analysis of cells on a microfluidic platform, specifically related to the quantification and monitoring of cells in a static solution, on-chip patch-clamp measurements, and examination of flowing cells. Another interesting application is presented in [9], where a biosensor working on both MOS capacitor and electric double-layer capacitor mode performs label-free DNA detection in real-time.

Since the aim of the work presented here is the examination and development of RF- or microwave-based sensors, a further look at the possibilities in this frequency range will be given. RF- or microwave-based sensing consists in exploiting electromagnetic waves in the frequency range from about 1–100 GHz and it is very promising because of the following properties. First, on the contrary to DC-electrical or mechanical techniques [8, 10], no contact between the sensor and the biomolecules or cells is indeed required, which suppresses the possible contamination or perturbation of the bio-material. Second, optical and chemical analysis require the addition of chemical substances or particles [11] which may modify the MUT and are time consuming. In the case of RF- or microwave-based sensors no markers are necessary. Third, the sensors are able to achieve relative deep penetration (tens of centimeters) in bio-matter (body for example), conversely to Terahertz or optical techniques, of course accompanied with less resolution. Fourth, the radiation produced by these sensors is non-ionizing, which means that it does not carry enough energy to alter the molecular structure of the MUT. Fifth, organic material has a unique frequency behavior that can deliver information about their structure and function. RF- or microwave sensors are able to extract these spectral signatures with a spatial distribution that permits the accurate test of several MUT simultaneously.

RF or microwave sensors consequently represent a high-risk/high reward technique for biosensing. High-risk since the biomedical community has to be deeply convinced to migrate from classical tools to microwave ones. High reward since combined with micro- and nano-technologies, the technique is highly parallelizable, and then constitutes a ground breaking solution for non-invasive high throughput screening, for which the market is estimated to reach $19.9 Billion by 2017 [12] as it is the central tool for new drug discovery.

The fundamental investigations on RF-sensing of biological media, mainly organs and tissues, were strongly initialized by Herman and Schwan during the late 1980–1990s, with a remarkable contribution of Gabriel who demonstrated and modeled different electrical signatures for various biological samples [13]. In the late 2000s the techniques received an impressive rising of interest as the previously mentioned abilities fit perfectly to achieve the non-invasive detection of tumorous zones inside the body. Khan et al. indeed point out that a strong contrast between tumorous and non tumorous tissues [14] can be reached thanks to microwave analysis.

One of the most important investigations with a microwave detection technique focuses on breast cancer imaging. It has taken place in USA and Canada (Calgary Univ. and Wisconsin Univ. notably) for the last decade [15, 16]. The resulting microwave-based imaging systems have reached maturity, and the one developed at the University of Bristol is already undergoing clinical trials [17–19]. The system components involves sophisticated sensor design and a complex RF switching network connecting the sensor array to a high-frequency imaging system. The experimental implementation and verification of these techniques in medical imaging is still ongoing. Based on these demonstrations at macro scale (organs, tissues), recent attempts to demonstrate the integration of the microwave sensing technique thanks to the micro- and nano-technologies are available [3, 20, 21]. The miniaturization of microwave systems is visible in daily life: cell phones and micro technologies demonstrated their abilities to revolutionize system architectures and performances [22]. As example, in [20] presents the detection of hybridized nucleic acids (DNA) on top of an inductor and a capacitor with silver functionalized particles to amplify the detection. Even if the placement of the biological sample is not appropriate for calibration and thus accurate quantification, the proof of concept is well established. Potentialities of such miniature microsystems are broad especially for high throughput screening applications.

For the specific application presented in this work, it is necessary to see the origins. The first reports of broadband dielectric spectroscopy-based micro-sensor, operating at microwave frequencies, within microfluidic channels for bio-analysis were presented in [5, 23, 24]. The device is composed of a microwave circuit elaborated on top of a quartz wafer, on which microfluidic circuitry is in charge of the processing of bio-materials and the location of the liquid under test over the microwave circuit to promote the electromagnetic field/liquid interaction. Broadband characterization of the microwave circuit is then performed and relative permittivity of the fluid is consequently extracted. Measurement results were presented of the permittivity of aqueous solution of Albumin over three frequencies-decades from 40 MHz to 40 GHz. It was demonstrated for the first time, the microwave-sensing of living cells proliferation in microsystems [24]. In this work a further implementation will be presented, where a metamaterial periodic structure will be used to make a similar analysis but instead of sensing one single sample of the MUT, several samples can be screen simultaneously with one single device and one read-out signal.

5.1.1 Frequency Characterization of Biological Material

The biological effects of electromagnetic fields on living tissue have been studied for over 100 years, since the discovery of electromagnetic radiation and its first use in therapeutic applications. The physical interaction between electromagnetic waves and biological systems are defined by maxwell's equations describing its characteristics and behavior. The Maxwell's equation along with the relative permittivity and permeability equations were already introduced in Sect. 2.1. Living matter generally have μ_r equal to that of free space, with exception of cells, molecules, or organisms endowed with ferromagnetic particles. On the other hand, ε_r shows a frequency dependency and it is also related to the medium where the living matter is located [25].

The dielectric properties of biological tissue provide information about the biophysical interaction of all the elements involved in a system and hence information about their structure and function. For the analysis of biological tissue the frequency spectrum is typically divided into three main dispersion regions, although other minor dispersion regions can also occur. First, the alpha dispersion region (10 Hz to 1 KHz) is associated with tissue interfaces such as membranes. Second, the beta dispersion region (1 KHz to several MHz) is associated with the polarization of cellular membranes and protein and other organic macromolecules. And finally, the Gamma dispersion region (several GHz) is due to rotations and deformations of small single polar molecules or groups. Frequently, the solvent itself induces the Gamma dispersion due mainly to the polarization of water molecules. In the microwave frequency range the beta dispersion and the Gamma dispersion regions are included.

It is known that both loss factor $\tan\delta = \varepsilon''/\varepsilon'$ and relative permittivity ε' vary widely in different biological tissues depending on the frequency. Therefore, the permittivity can be understood through its relation to the bounding of the charges that produces a polarization of the material in response to the electric field such as the one described in the beta dispersion region. This is due to the fact that the electrical conduction in biological tissues, unlike metallic conductors, is due to ions. Another feature to acknowledge is the high losses in the GHz regime which originate from the relaxation of water inside the material under test.

The dielectric properties of materials are characterized by the measured relative complex permittivity ε and was shown in Eq. (2.6). The real part is defined as the dielectric constant and the imaginary part is the measure of electrical energy dissipation in terms of dielectric losses. Therefore ε'' is related to conductivity as follows [26]

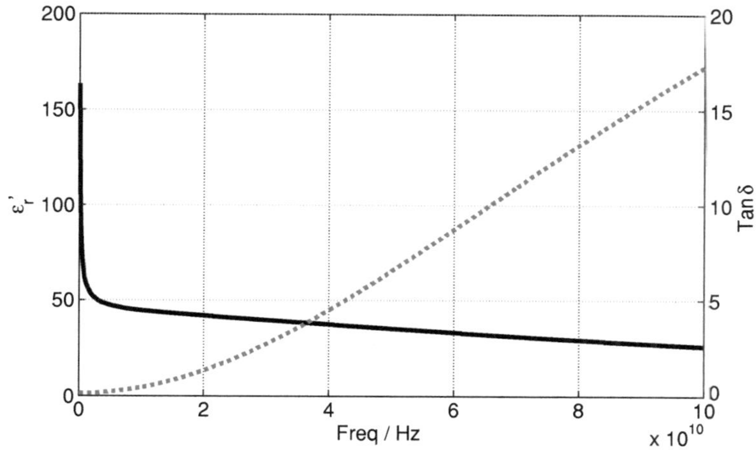

Fig. 5.1 Dielectric permittivity and electrical conductivity of liver as a function of frequency [13]

$$\varepsilon'' = \frac{\sigma}{\omega\varepsilon_o} \tag{5.1}$$

where σ is the conductivity, ε_o is the permittivity of the free space and ω is the angular frequency of the field. Depending on the nature of the material, a contribution from the frequency independent ionic conductivity can be also present. An example is depicted in Fig. 5.1 for liver tissue where the Cole–Cole model was used. All of the living tissues show a similar trend to the liver, of course with different values of permittivity and conductivity but in general can be said that biological materials exhibit very high dielectric constants, specially at lower frequencies. This is because they are composed of macromolecules, cells and other membrane-bound substances. At low frequencies the dielectric constant is dominated by the membrane capacitance of the cell which is very high and it undergoes dramatic changes as the frequency increases slightly (<3 kHz). Charges start to accumulate at the boundaries of different materials such as intra and extra cellular spaces due to the applied electric field. Above 1 kHz, the capacitance becomes short-circuited facilitating the participation of intracellular fluid in electric current conduction. It is not clear if the reason for the capacitance increase is due to changes in the membrane area and thickness or if other factors are included in this process [27]. This causes the conductivity to increase with increasing frequency. As frequency increases to higher values (10 kHz to 30 MHz), there is not enough time in each cycle to completely charge the cell membrane and therefore, the total charge per cycle decreases, along with the membrane capacitance and as a consequence the dielectric constant. For higher frequencies (>30 MHz), the charge in the membrane stabilizes until the rotational and vibrational properties of polar molecules of water become significant and the rotation of these molecules produces a high loss that results in a very high conductivity [25]. The main features of the dielectric spectrum of tissues have been reported already by Foster and Schwan in [28]. As mention already, the dielectric spectrum of tissues have three main relaxation regions, each representing the manifestation of a polarization mechanism characterized by a single time constant, τ which, as a first approximation modeled by the Debye equation represents the complex relative permittivity as follows [13]

$$\varepsilon(\omega) = \varepsilon_\infty + \frac{\varepsilon_s - \varepsilon_\infty}{1 + j\omega\tau}. \tag{5.2}$$

In the above equation ε_s corresponds to the low frequency limit ($\omega\tau \ll 1$) and ε_∞ the high frequency limit ($\omega\tau \gg 1$). The frequency of the applied EM field is represented by ω and τ is the relaxation time. When separated in real and imaginary part the $\varepsilon(\omega)$ looks as follows

$$\varepsilon'(\omega) = \varepsilon_\infty + \frac{\varepsilon_s - \varepsilon_\infty}{1 + j\omega^2\tau^2} \tag{5.3}$$

$$\varepsilon''(\omega) = \frac{(\varepsilon_s - \varepsilon_\infty)\omega\tau}{1 + j\omega^2\tau^2}. \tag{5.4}$$

Since tissues are highly complex inhomogeneous materials and exhibit different dispersion regions, these multiple dispersion regions can be accounted by enhancing the Debye equation with an additional parameter α that represents the measure of the broadening of the dispersion. This new equation is known as the Cole–Cole equation

$$\varepsilon(\omega) = \varepsilon_\infty + \frac{\varepsilon_s - \varepsilon_\infty}{(1 + j\omega\tau)^{(1-\alpha)}}. \tag{5.5}$$

The generalized form of the above equation is expressed as follows

$$\varepsilon(\omega) = \varepsilon_\infty + \Sigma_n \frac{\varepsilon_s - \varepsilon_\infty}{(1 + j\omega\tau_n)^(1 - \alpha_n)} + \frac{\sigma_i}{j\omega\varepsilon_o}. \tag{5.6}$$

By changing the parameters for different types of tissues the above equation can model its permittivity with respect to frequency. Conductivity is also an important parameter which can be modeled by Debye relaxation equation for different types of tissue with the equation

$$\sigma_m = \frac{\varepsilon_0(\varepsilon_s - \varepsilon_\infty)\omega^2\tau}{1 + \omega^2\tau^2}. \tag{5.7}$$

Another parameter to account for is called skin depth, which explains the propagation characteristics in the medium where the biological material is located. The skin depth is defined as the point inside the tissue where the EM field strength is attenuated to $1/e$ of the initial value. This is described as

$$\delta_m = \frac{1}{\sqrt{\pi f \mu \sigma_m}} \tag{5.8}$$

where f is the frequency and μ is the permeability of the medium. At low frequencies the permittivity is relatively high and conductivity is low, and fields can penetrate much deeper without strong attenuation. As the frequency is increased the skin depth is decreased because of higher losses in the tissue.

Another parameter which has been reported in recent studies is temperature dependency of the dielectric properties of tissues [29]. The study is done using a single-pole Cole–Cole model with second order polynomial parameter to add temperature effects. It was demonstrated that is not possible to consider a linear temperature coefficient to explain the change in the dielectric properties due to the change in temperature. On the other hand the effects during a heating procedure are very different than when the tissue is being cooled. For example, when the temperature of the tissues reaches more than 46 °C irreversible damage to the cells occur and the increase in temperature can evaporates the water contents in the tissue, which leads to variations in the dielectric properties. This will be addressed in the next chapter where a sensor that performs thermal ablation will be presented.

5.2 Sensor Design and Fabrication

The main application of a planar microwave sensor in the biological field is for analysis of different properties of cells. Different information can be extracted from the dielectric properties of the cells, for example, concentration, state of the cell in their living cycle, response to exposure to drugs. The goal was to design CRLH-TL structures with optimized properties in terms of size and sensitivity that could be integrated with microfluidic technology. The boundary conditions for the optimization are given by the availability of the used fabrication technology, operation frequency and the measurement hardware. The initial idea is depicted in Fig. 5.2. The sensor will consist of a CRHL-TL line with n_c unit cells, because then the MRPT presented in Sect. 3.1 can be used for the data extraction. One capacitor in each unit cell is electromagnetically coupled to a microfluidic channel. Hence the liquid in the channels will influence the capacitance value and with it the transmission line properties of the complete CRLH-TL. By measurement of the scattering parameters of the sensor the dielectric properties of the liquids in the microfluidic channels can be determined one by one and hence monitored simultaneously. The design of the sensor is divided into two phases: First, the determination of equivalent circuit parameters and second the transfer of the equivalent circuit into a real layout. The available model for the design was refined in terms of accuracy and input parameters. The parameters included are given in Table 5.1.

The reason for the chosen values of the input parameters will be addressed. The center operation frequency was selected to be around 30 GHz since at this frequency the Gamma dispersion region is valid and most of the desired properties to be studied

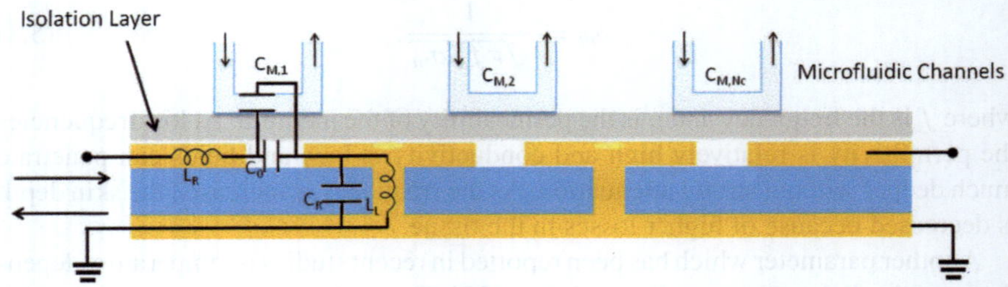

Fig. 5.2 Sensor equivalent circuit for three unit cells and their correspondent microfluidic channels [30]

Table 5.1 Parameters for the design of a CRLH-TL sensor with microfluidic channels for analysis of biological cells

Parameter	Desired value
Center operation frequency	30 GHz
Bandwidth	15 GHz
Sensitivity $(\Delta\varepsilon/\varepsilon)_{min}$	5 %
Number of microfluidic channels	4
Dynamic range $(\Delta\omega_r/\omega_r)_{min}$	10^{-3}

in the cells are somehow related to the polarization of its water molecules. The bandwidth is given to include a broad range of frequencies where several unit cells can be included as sensing elements. The desired sensitivity is a conservative value to be able to extract slight changes in the properties of the cells. The number of microfluidic channels should not be higher than four for initial prototypes since the fabrication of a larger structure brings severe difficulties especially to accommodate the microfluidic channels. Furthermore, the calibration measurements required in the MRPM of a larger structure would be very problematic until the filling and emptying of the channels can be done in an automatic way. Finally, the dynamic range was selected around 100 MHz to be able to detect the small shifts in the resonant peaks produced by the MUT, since the information that needs to be extracted does not create a high change in the dielectric properties of the cells. For example, a comparison between a pure cell culture and a cell culture exposed to a particular drug will result in a dielectric change of no more than $\varepsilon_r = 5$.

The presented model of Fig. 5.2 is a periodical structure where six parameters are sufficient to define it. In each unit cell one capacitor can be defined as

$$C_{L,n} = C_0 + C_{M,n} \tag{5.9}$$

where

$$C_{M,n} \approx \varepsilon + \Delta\varepsilon_n. \tag{5.10}$$

The capacitor $C_{L,n}$ represents the electromagnetic coupling between the structure and the adjacent microfluidic channel. Hence, the dielectric constant of the liquid in the channel ($\varepsilon + \Delta\varepsilon_n$) influences the capacitance value $C_{L,n}$. The sensitive part of $C_{L,n}$ is labeled $C_{M,n}$ and the fixed part C_0. The ratio between C_M and C_0 is controlled by the geometrical placing of the microfluidic channel and the thickness of an isolation layer. The reason for using an isolation layer between the CRLH-TL and the microfluidic channels is not only the high dielectric losses inherent to all biological material but more important the high conductivity, without this layer the sensor would be short-circuit and therefore not perform correctly. Since the isolation layer has a low conductivity it prevents the currents from the sensor structure to interact with the lossy tissue and therefore the signal in the sensor is not degradated. With the help of the MRPM the dielectric loading of the microfluidic channels from Fig. 5.2 $\left[\Delta\varepsilon_1\ \Delta\varepsilon_2\ \ldots\ \Delta\varepsilon_{n_c}\right]^T$ can be derived based on a measurement of the CRLH TL properties. In order to improve this model we are aiming towards the inclusion of parameters from the physical layout e.g. the thickness of the isolation layer, channel width, etc. This step will simplify the transformation of the equivalent circuit to the final layout and limit the number of optimization steps in order to achieve a desired sensor performance.

The MRPM presented in Sect. 3.1 will be used for extracting the dielectric profiles. For this method the sensor should operate in the LH band of the CRLH-TL because it is beneficial in terms of bandwidth requirements [31] and it is sufficient for the appropriate extraction. Since it is not a linear system we use an iterative process based on the MRPT to find solutions for the parameters C_L, L_L, C_R, L_R and the ratio between the fixed and the variable capacitance C_0 and C_M. Generally, there exist several solutions, but not all of them are realizable with a certain fabrication technology. The ratio C_M/C_0 has been chosen with respect to the requirements on sensing accuracy, system parameters and especially the losses of the liquid under test. An example of the influence of this ratio to the maximal number of measurement channels for a typical system configuration is shown in [30]. For this particular case it was found that a maximum of 18 cells can be obtained for a ratio of $C_M/C_0 = 3$. For smaller values the limitation comes from the frequency resolution for the detection of shifts in resonance peaks and for larger values the losses due to the liquid under test limit the number of channels n_C. It can also be observed the influence of ohmic losses in the reactive line elements 1 dB/cell, the change in the curve if the liquid has lower loss ($\tan\delta = 0.7$) and the influence of a change in $(\Delta\varepsilon/\varepsilon)_{min}$ from 5 to 3 %.

In the second step the most suitable solution must be transfer into a layout, verified and optimized with full wave electromagnetic simulation. CST Microwave Studio is used to simulate a sensor design using a coplanar waveguide (CPW). The reason for choosing a CPW design for the sensor is because it exhibits a high interaction between the structure and the MUT placed on top of the structure since the field can protrude far into the space above the CPW line. Another reason is that CPW is well suited for broadband applications and can be easily integrated with periodic metamaterial structures.

In Fig. 5.3 is shown the design for a two unit cell structure using the software CST Microwave Studio. The structure is made of interdigital capacitors with length

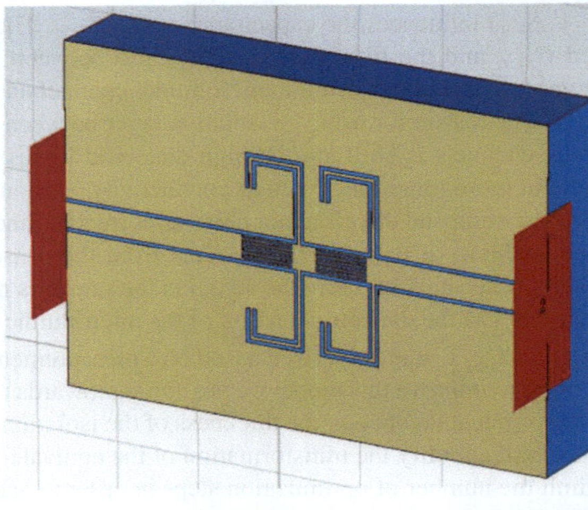

Fig. 5.3 Design of the sensor structure in CST Microwave Studio. The structure is made of interdigital capacitors with length 300 μm and gap between the fingers of 10 μm and spiral inductances with a total length of 920 μm

Fig. 5.4 Dispersion diagram of the structure depicted in Fig. 5.3. The LH and RH band are depicted as well as the band gap

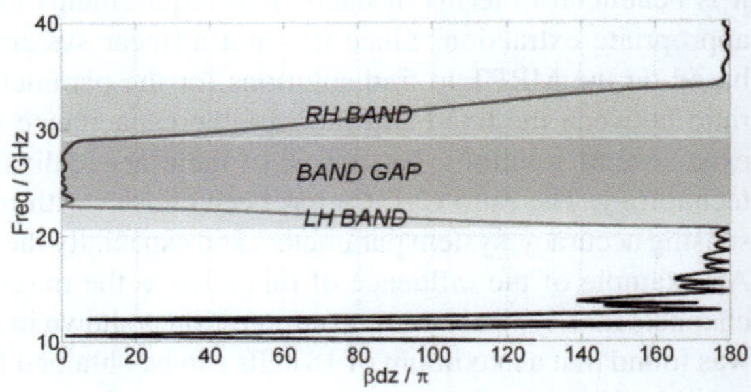

300 μm, gaps between the fingers of 0 μm and spiral inductances with a total length of 920 μm. The substrate used is Quartz with a thickness of 500 μm. To compare with the ideal discrete structures presented in Sect. 2.2.2, the dispersion diagram is shown in Fig. 5.4. The RH band and LH band can be seen as well as the band gap between them. The frequency of the two bands is as expected between 20 and 35 GHz. The ripples shown under 20 GHz most probably represent the long feeding CPW line necessary to have enough room for placing the complicated microfluidic network. To corroborate the results, α and β are shown in Fig. 5.5. Here can be seen again the two transmission bands and the already mentioned ripples under 20 GHz.

The sensing element of the shown design corresponds to the interdigital capacitor that is represented by $C_{L,n}$ in Eq. (5.9) and corresponds to $C_0 + C_{M,n}$ in the the equivalent circuit depicted in Fig. 5.2. This capacitor corresponds to the C_{LH} of a typical CRLH-TL. The reason to choose this specific capacitor as the sensing element is the higher sensitivity that it features in comparison with the C_{RH} but that

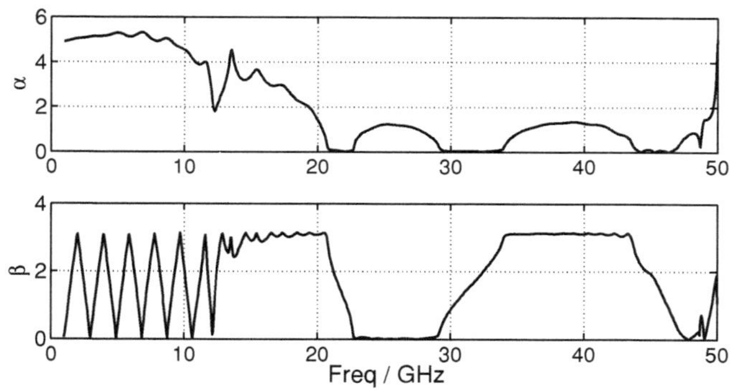

Fig. 5.5 α and β of the structure depicted in Fig. 5.3. The LH and RH band are depicted as well as the band gap

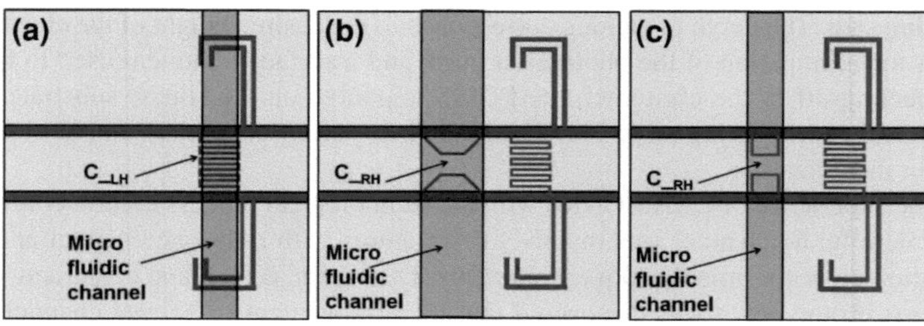

Fig. 5.6 Different unit cells used in the design of the prototype sensors. **a** The sensing element is the C_{LH}. **b** and **c** The sensing element is the C_{RH}

also means that it is more susceptible to the high losses inherent from the biological tissue being analyzed. For this reason additional structures were designed with the C_{RH} capacitance used as sensing element, from now on referred as right-handed sensors, to check if although their sensitivity is expected to be lower, maybe it is sufficient. The different unit cells that were used in the final designs are depicted in Fig. 5.6. In Fig. 5.6a is the already discussed structure where the interdigital capacitor i.e. the C_{LH} is the sensing element or the so- called left-handed sensor. In Fig. 5.6b, c are the two geometries used in the right-handed sensors since from the location of the microfluidic channel it is clear that they use the C_{RH} as sensing element.

For the final designs, all the unit cell layouts were used in configurations containing one, two and four unit cells. In the following section the fabrication process and the challenges faced in this stage will be discussed.

5.2.1 Technology Process

The fabrication of these structures was done in close cooperation with LAAS-CNRS in Toulouse, France since they have a wide experience in microfluidic technology combined with microwave devices. The procedure for all the designed structures is the same and consists of the following steps. After cleaning the quartz substrates a photoresist is spin-coated and patterned using UV photolithography. The wafer is then placed into an evaporating system, which enables the deposition of a seed layer of titanium followed by 0.3 μm thick gold layer. The metal is exposed during the elimination of the photoresist in a solvent bath of acetone. The fluidic part which is realized in polydimethylsiloxane (PDMS) may be prepared in parallel. The elaboration of the fluidic channels is performed through the replication of the elastomer in a silicon mould. This one is previously prepared with the pattern of a photoresist, which is used as a mask during the etching of a silicon substrate with a Reactive Ion Etching step. The depth of etching corresponds to the desired height of the channels. After the elimination of the photoresist mask and a surface treatment used to favor the peeling-off of the elastomer, the PDMS is poured on the silicon substrate and solidified with a curing step. The PDMS is then peeled off and assembled on the circuit metallization.

The experience of LAAS-CNRS with the technological process used to construct the microfluidic channel was mainly for structures with only one channel and no isolation layer for broadband measurements. During the design and construction of the first prototypes it was discovered that the arrangement of several channels on one structure is quite difficult. To position the entrances and exits of the channels was challenging. Additionally, the channels are too close together and that translates on them being not truly independent from each other. The under-pressure applied on one channel can affect its neighbors and as a result they can detach from the structure and leakage can occur. And finally, the sensor requires an isolation layer to prevent the degradation of the signal due to the high conductivity and inherent losses of the MUT. Unfortunately, its thickness creates a mechanical problem for proper gluing the PDMS channels to the structure resulting again in an undesired leakage of the MUT.

As ongoing work, a dedicated test bed which allows the electrical contacting of the sensors and at the same time the automated filling of the channels is being realized. During the tests of the first planar sensor prototypes the liquids were indeed injected using syringes manually actuated as seen in Fig. 5.7. Although this procedure was satisfactory for devices classically developed at LAAS-CNRS, the first tests of the prototypes presented in this work demonstrated that a more precise control of liquid injection is required (consequently to the low size—low volume—of the fluidic channels). A dedicated setup including solenoid valve and associated control electronics will be specifically developed to assure a precise fluidic control during the further experiments of these structures within the framework of a joint project with the LAAS-CNRS team.

Fig. 5.7 Microwave and fluidic actual test setup (manual fluidic control)

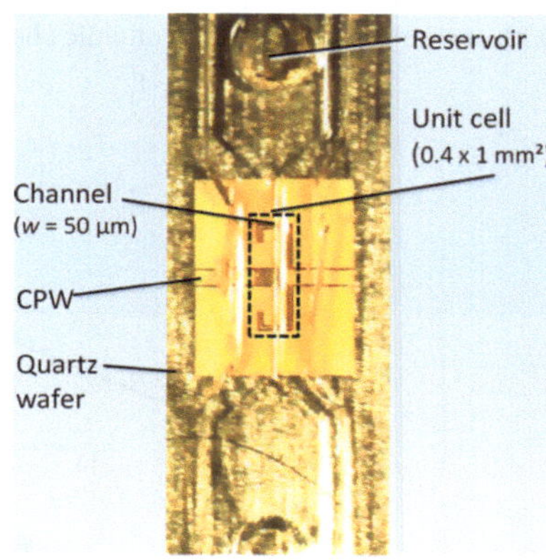

Fig. 5.8 Developed prototype of a planar structures for analysis of biological cells. The microfluidic reservoir used for filling of the channel is depicted as well as the 50 μm microfluidic channel [30]

5.2.2 Prototypes and Measurements

The examples of the first planar sensor prototypes that were developed in cooperation with LAAS-CNRS for the analysis of dielectric properties of organic cells are depicted in Figs. 5.8 and 5.9. In Fig. 5.8 is visible the microfluidic reservoir used for filling the channel and the channel itself with a width of 50 μm. The realization of the lumped element inductor L_{LH} and capacitor C_{LH} with a CPW stub line and an interdigital capacitor is also depicted. The size of the unit cell is approximately $0.4 \times 1\ \text{mm}^2$. For the control of the value C_M/C_0 a 4 μm thick SiO_2 layer has been inserted between the metallization and the channels. The structures are made of a periodical arrangement of identical unit cells and as mentioned previously, several different prototypes were constructed and they can be grouped in two categories. The first category is the RH sensors with C_{RH} as sensing element as shown in Fig. 5.9a, b. The second category is the LH sensors where the C_{LH} is the transducing element as shown in Fig. 5.9c. Structures were developed with one, two and four unit cells.

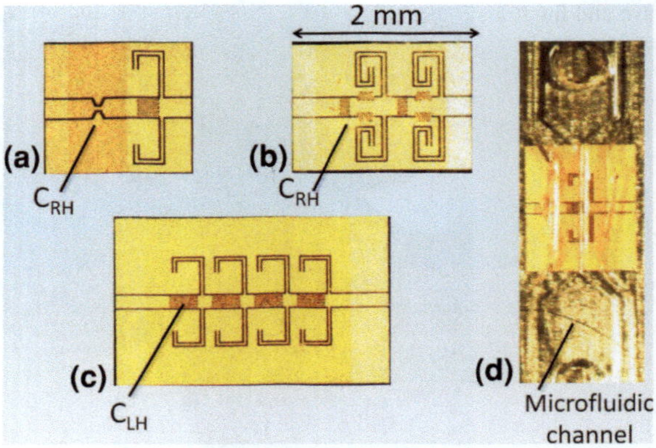

Fig. 5.9 Developed prototypes of a planar structure for analysis of biological cells. **a** RH sensor with one unit cell, **b** RH sensor with two unit cells, **c** LH sensor with four unit cells, **d** LH sensor with one unit cell and the microfluidic channel system mounted on

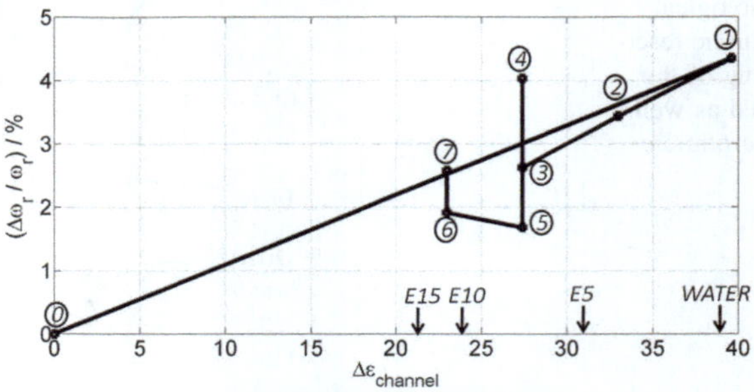

	Ethanol	Water
E5	5%	95%
E10	10%	90%
E15	15%	85%

Fig. 5.10 Measurement of relative frequency shift versus varying dielectric constant in the microfluidic channel $\varepsilon_{r,Fluid}$. The structure is a LH sensor with one unit cell fabricated in the first run [30]

The structures with one unit cell were constructed to test the behavior of the different geometries and to decide which one had a better performance in terms of accuracy. As a result, the structures with LH capacitors as sensitive elements have higher sensitivity but hence, are more susceptible to high losses from the MUT and the ones with RH capacitances are vice versa. The structures with two unit cells serve the purpose of analyzing the extraction of a profile by using the MRPM and finally the structures with four unit cells are used to prove the complete goal of the project that is to test simultaneously several MUTs and extract a complete dielectric profile. On top of the structures microfluidic channels made of PDMS were mounted as depicted in Fig. 5.9d.

Measurements to prove the concept were done by filling the channels with water and different concentrations of a water and ethanol mixture. To introduce the MUT into the channel, a small droplet was placed on one of the holes of the PDMS over the reservoirs. By applying under-pressure on the exit hole the liquid is taken in by the channel. Figure 5.10 shows the representative results for the characterization of a device with one channel. The idea is to fill the channel successively with liquids with different dielectric constants and to monitor the frequency shift. In case everything is correct, it was expected a linear relation between the dielectric of the liquid and the relative change of resonance frequency if the changes are less than 10 %. All measurements are depicted in Fig. 5.10 and the procedure started with the determination of the unloaded resonance frequency by an empty channel measurement and is marked with the number zero. Then, a measurement of the largest frequency shift was done by using the liquid with the highest dielectric constant (water in this case) which is marked with the number one. Finally, measurements of interim values using liquids with lower dielectric constant made of mixtures of water and ethanol at 5, 10, 15 % were performed and are marked as two to seven. Between the measurements one to seven the channels were emptied and the unloaded resonance frequency was measured.

The graph reveals problems that were already observed with several of these first prototype structures. Following the measurement procedure described above for one liquid under test e.g. 10 % ethanol and 90 % water (e10), different values for the frequency shifts are obtained. The too low values can be explained with problems associated with filling of the 50 μm channels since they are very narrow and it is difficult to assess if they are homogeneously filled. On the other hand, the too high values are supposed to originate from bad adhesion of the channels and therefore leakage of the MUT into the structure. The liquid under test can penetrate into the structure between the PDMS film and the metallization. This assumption is confirmed by observations with the microscope during the filling of the channels.

To correct the mentioned problems of the structures a new group of prototypes with the same characteristics and design were constructed with some improvement in the technological process. First, the previous isolation layer realized in the polymer SU8 has been replaced by a mineral dielectric of silicon nitride deposited through low plasma enhanced chemical vapor deposition (PECVD). Its higher dielectric permittivity permits to lower its thickness to 0.25 μm. It therefore contributes to decrease leakages of the fluidic channels. This second and major modification, which fully solves any risk of leakage, corresponds to the microfabrication of the channels in SU8 instead of the elastomer PDMS. PDMS presents indeed the drawback to be manually assembled to the metallized substrate with a limited adhesion on the surface, whereas the SU8 polymer requires conventional and collective photolithographic steps and provides a strong attachment to the lower wafer. It consequently avoid any leakage below the channels and any mixing of the analyzed solutions.

The results of the same structure with one channel, but with the new technology to construct it, is shown in Fig. 5.11. The measured liquids were again pure water and mixtures of water and ethanol at 5, 10, 15, 20, 50, 100 %. The extracted values are within tolerance limits of the theoretical values and the drift observed with the

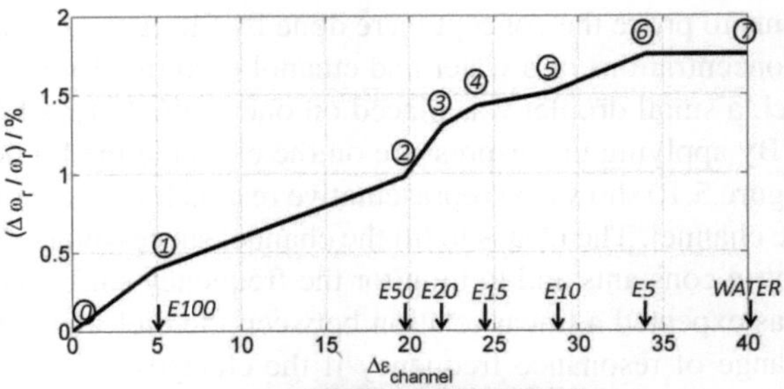

	Ethanol	Water
E5	5%	95%
E10	10%	90%
E15	15%	85%
E20	20%	80%
E50	50%	50%
E100	100%	0%

Fig. 5.11 Measurement of relative frequency shift versus varying dielectric constant $\varepsilon_{r,Fluid}$ in the microfluidic channel for the second group of developed prototypes. The structure is an LH sensor with one unit cell

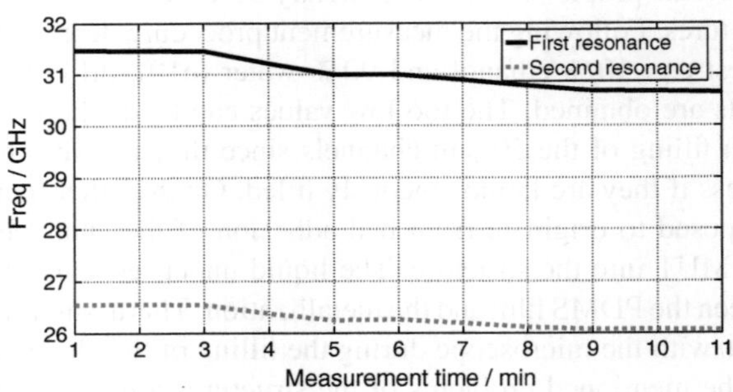

Fig. 5.12 Stability measurements of the LH sensor with one unit cell

previous prototype is no longer visible, since the problem with the leakage was solved. Also from this figure, it can be observed that to make a successful extraction a high resolution is necessary. The maximum relative frequency change between an empty measurement and a filled measurement with pure water (highest possible dielectric value) is less than 2 %. Therefore, the dynamic range of 10^{-3} was an important criterion in the design.

Since the filling of larger structures can take a longer time, specially due to all the calibration measurements that are necessary, it is important to analyze what happens when the channel is filled for a large period of time. An experiment was done by filling a one channel structure with water and ethanol at 15 %. The measurements were taken every minute for a period of 10 min and the results are shown in Fig. 5.12. For the first three minutes the frequency values remain stable and after this period of

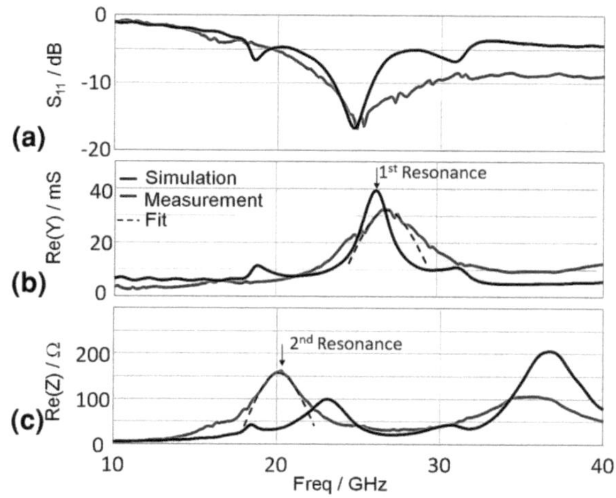

Fig. 5.13 Example for **a** the S-, **b** Y- and **c** Z-parameters of a two cell LH sensor, where both channels are filled with water [30]

time a slight drift begin to occur most probably due to the fact that the ethanol starts to evaporate. From these results it can be deducted that a complete measurement should be carried in less than three minutes and a more efficient filling method needs to be introduced for larger structures.

Additional measurements with two channel structures were carried out successfully with the structures designed in the second run where the leakage of the channel was already solved. Figure 5.13a shows the measured input reflection coefficient S_{11} of the structure from Fig. 5.8b from measurement and simulation. The main peak at 25 GHz is similar for both curves but there is a difference in the resonance around 19 GHz for the simulated and 17 GHz for the measured data. Additionally, the measured losses in the structure are significantly higher above 20 GHz. This discrepancy is attributed to a deviation of the properties of the used materials like the evaporated gold and PDMS from their ideal behavior. However, since the elements of the matrix A are determined from calibration measurements this should have no influence on the performance of the sensor.

To be able to apply the MRPM, the measured S-parameter data has to be then transformed to Z- and Y- parameters. Figure 5.13b, c show the plot of their real parts, which are used for the detection of resonances. In this case the resonator is analyzed as a one port open-ended device. Hence the first resonance can be detected with a maximum in the real part of the admittance Y, for the detection of the second resonance the Z- parameters are employed. Further resonances can be detected, if necessary, by the alternating analysis of Z- and Y-parameters. In order to increase the accuracy of detection the actual resonance frequencies are determined by a local curve fitting procedure around the maxima. The corresponding fitting curves are depicted with dashed lines in Fig. 5.13b, c.

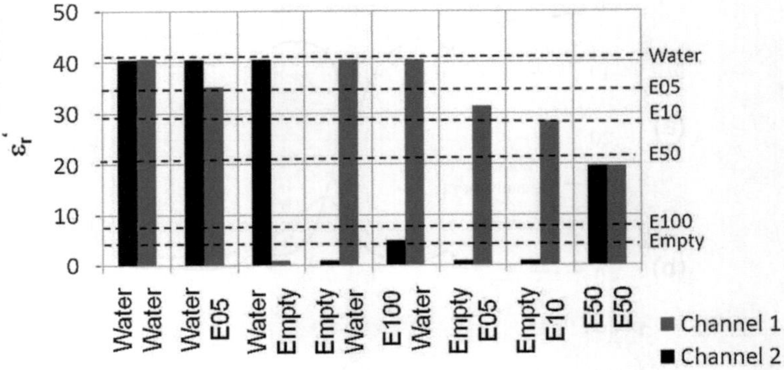

Fig. 5.14 Examples for the monitoring of two microfluidic channels using the MRPT. The *horizontal axis* of the figure lists the liquids injected into the four channels, whereas the *vertical axis* shows the extracted values of the dielectric constant [30]

Figure 5.14 shows some examples for the monitoring of sensors with two microfluidic channels using the MRPT. Prior to the measurement, the determination of the matrix A was of course performed through calibration measurements by applying specific filling patterns with water-filled and empty channels. Different combinations and sensors have been tested. The observed accuracy is within what was expected. In the first two measurements, pure water content in both channels and the difference between pure water and water with 5 % ethanol content (E05) were successfully extracted with the MRPM. The following two measurements detect empty versus water, which means that when two adjacent channels are filled with very different materials the extraction is still possible. The last four measurements are done with different profiles where combinations of empty, water and water-ethanol mixture of different concentrations are inserted in the channels. The profiles were successfully extracted, therefore it can be concluded that the results are reproducible.

In Fig. 5.15 are depicted examples for the monitoring of sensors with four microfluidic channels using the MRPM. The test is done with a structure that uses the unit cell shown in Fig. 5.6a, where the C_{LH} is the sensing element in a four channel configuration. The figure shows four different profiles with the channels filled either with water or different concentrations of water-ethanol. The first thing to consider is that the calibration measurements were very difficult to make and it is not certain if the channels were appropriately filled for all these measurements. For such large structures an automatic setup for filling the channel is absolutely necessary. As for the profiles they all have errors in the extracted values most probably coming from the errors in the A matrix itself or again because the filling of the channel was not done correctly.

Although it is not possible to quantitatively compare the LH sensors from the RH ones, a few conclusions can be drawn from the experiments made. The LH sensors are definitively more sensitive when loaded by a MUT. For these sensors, the thick isolation layer is indispensable and creates a challenge in the fabrication process to be

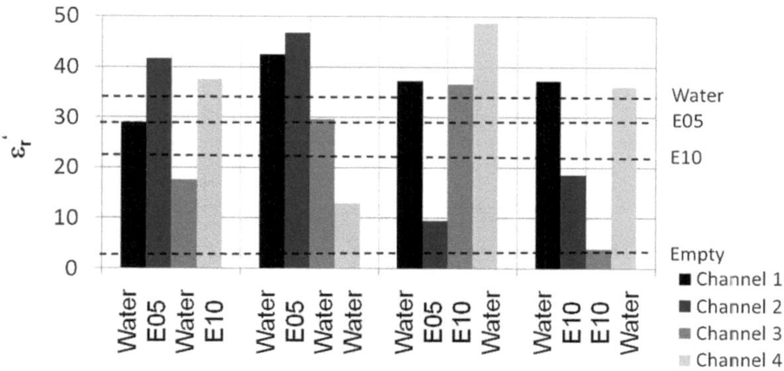

Fig. 5.15 Examples for the monitoring of four microfluidic channels using the MRPT. The *horizontal axis* of the figure lists the liquids injected into the four channels, whereas the *vertical axis* shows the extracted values of the dielectric constant

able to adequately bind the microfluidic channel to the structure itself. On the other hand, this sensitivity is put to good use for the extraction of the profiles, they are more accurate than the profiles from RH sensors but when the dielectric constant of the liquids used in the calibration is very different, than the one from the MUT, problems arise in the extraction procedure. The RH sensors are of course the opposite. They have lower sensitivity, but can function with a thinner isolation layer, in other words, they can be a little more robust but the profiles are less accurate.

5.3 Analysis of Biological Cells with Microwave Sensors

There are many important types of information that can be extracted from the dielectric properties of biological cells. For example, all cells have a reproductive cycle that includes four very defined stages, each with a unique dielectric signature. By monitoring the changes in dielectric properties it is possible to identify at what stage of the cycle the cells are currently, and further information on concentration, presence of certain compounds, such as proteins or others, would also be available. On the other hand, cancer cells accumulate cell cycle alterations and abandon cell cycle control making a completely different dielectric image as normal cells, therefore, this is an interesting field where monitoring of dielectric properties could play an important role.

With the presented sensors, measurements of biological cell pools were carried out to analyze the dielectric behavior of cells and its related biophysical and physiological information. The living cells employed in this study come from a chinese hamster ovary (CHO) cells line, which constitute a well-known mammalian cell model for biological and medical research. They are traditionally used for in vitro investigations related to genetics, toxicity screening and gene expression notably. The cells are

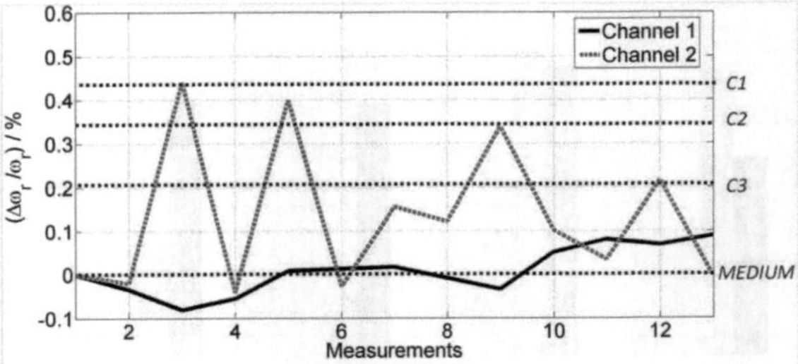

Name	Percentage of cells (million / ml)
C1	100
C2	50
C3	25

Fig. 5.16 Biological cell measurements. The structure used is a RH sensor with two unit cells. Different concentrations of chinese hamster ovary (CHO) cells can be seen

grown in regular Petri dish and then placed in suspension in their culture medium, which consists in Minimum Eagle's Medium (MEM) supplemented with 8 % fetal calf serum, 1 % peniciline and 1 % streptomycine.

The measurement results are depicted in Fig. 5.16 with different concentrations of cells in the MEM. The structure that was used for these measurements has two channels and used the C_{RH} as a sensing element. There are 12 different measurements depicted. The first channel is always filled with the pure culture medium, a gel designed to support the growth of microorganisms or cells, and is depicted in the figure with the solid black line. On the second channel, different concentrations of cells suspended in the culture medium are injected and empty measurements between the different profiles are done to ensure that the channel is truly empty. As for the cell concentration measurements, it is shown one measurement with concentration 100 million/ml, two measurements with concentration 50 million/ml and two measurements with concentration 25 million/ml. Although the absolute dielectric properties of the cells suspended in the culture medium was not possible to extract, the relative change is clear in the figure. Not only the sensor is able to discern between pure medium and medium with cells, but it also shows the different concentrations. The reason why it was not possible to detect the absolute dielectric value is because the calibration was done using pure medium and empty channels which has a very different dielectric value introducing a high error in the MRPM extraction. For maximum accuracy of the profile extraction, the MRPM requires to calibrate with a material of known dielectric properties and they have to be as close as possible to the actual MUT dielectric properties.

In summary, the proof of concept for a planar microwave sensor array with microfluidic channel for analysis of biological cells was successfully carried out. It was possible to detect the concentration of chinese hamster ovary cells in two fluidic channels independently by using the MRPM. The accuracy of the process for the structures with one and two fluidic channels was adequate and dielectric profiles were extracted. For future work, the devices are going to be used in an applicative case to screen chronic lymphocytic leukemia within the framework of a join project with the team from CNRS-LAAS. The goal is to achieve the same results as presented here, but with larger structures, where four or more fluidic channels can be monitored simultaneously. Furthermore, the designed sensors could eventually be integrated in a lab-on-chip environment thanks to their high possibility of miniaturization and appropriate performance.

References

1. R. Bashir, BioMEMS: state-of-the-art in detection. Opportunities and prospects. Adv. Drug Deliv. Rev. **56**(11), 1565 (2004)
2. E. Ghafar-Zadeh, M. Sawan, D. Therriault, A 0.18-m CMOS capacitive sensor lab-on-chip. Sen. Actuators A **141**, 454–462 (2008)
3. G. Facer, D. Notterman, L. Sohn, Dielectric spectroscopy for bioanalysis: from 40 Hz to 26.5 GHz in a microfabricated waveguide. APL **78**(7), 996–998 (2001)
4. K. Grenier, D. Dubuc, New broadband and contact less RF/microfluidic sensor dedicated to bioengineering, in *Proceedings of IEEE MTT-S International Microwave Symposium Digest*, (Boston, USA 2009)
5. K. Grenier, D. Dubuc, Integrated broadband microwave and microfluidic sensor dedicated to bioengineering. IEEE Trans. Microwave Theory Tech. **57**(12), 3246–3253 (2009)
6. H.-B. Liu, Sensing minute change in biological cell monolayers with THz differential time-domain spectroscopy. Biosens. Bioelectron. **22**, 1075–1080 (2007)
7. A. Treizebré, THz microscopic investigation on living cells, in *5th International Conference on Microtechnologies in Medicine and Biology*, QuTbec, Canada, April 2009
8. N. Bao, Recent advances in electric analysis of cells in microfluidic systems. Anal. Bioanal. Chem. **391**, 933–942 (2008)
9. Y.-B. Wang, A novel dual mode capacitor biosensor for real-time, label-free DNA detection, in *International Electron Devices Meeting*, December 2006
10. A. Han, Whole cell electrical impedance spectroscopy for studying ion channel activity, in *13th International Conference on Solid-State Sensors, Actuators and Microsystems*, (Seoul, Korea, June 2005)
11. J. El-Ali, P. Sorger, K. Jensen, Cells on chip. Nature **442**, 403–411 (2006)
12. GIA Report. http://www.strategyr.com
13. S. Gabriel, R.W. Lau, C. Gabriel, The dielectric properties of biological tissue: III. Parametric models for the dielectric spectrum of tissues. Phys. Med. Biol. **41**, 2271 (1996)
14. U. Khan, N. Al-Moayed, N. Nguyen, K. Korolev, M. Afsar, S. Naber, Broadband dielectric characterization of tumorous and nontumorous breast tissues. IEEE Trans. Microwave Theory Tech. **55**(12), 2887–2893 (2007)
15. C. Fear, Confocal microwave imaging for breast cancer detection: localization of tumors in three dimensions. IEEE Trans. Biomed. Eng. **49**(8), 812–822 (2002)
16. M. Lazebnik et al., A large-scale study of the ultra wideband microwave dielectric properties of normal breast tissue obtained from reduction surgeries. Phys. Med. Biol. **52**(10), 2637–2656 (2007)
17. M. Klemm, Experimental and clinical results of breast cancer detection using UWB microwave radar, in *Proceedings of IEEE Antennas and Propagation Society International Symposium* (2008)
18. M. Klemm, Clinical trials of a UWB imaging radar for breast cancer, in *Proceedings of 4th European Conference on Antennas and Propagation (EuCAP)*, (April 2010)

19. M. Klemm, Microwave radar-based differential breast cancer imaging: imaging in homogeneous breast phantoms and low contrast scenarios. IEEE Trans. Antennas Propag. **58**(7), 2337–2344 (2010)
20. N. Denef, RF detection of DNA based on CMOS inductive and capacitive sensors, in *IEEE European Microwave Week*, (Holland, Amsterdam 2004)
21. Y. Kim, Biosensors for Label Free Detection based on TF and MEMS Technology. Sens. Actuators, B: Chemical **119**, 592–599 (2006)
22. L. Katehi, Heterogeneous wafer-scale circuit architectures. *Microwave Magazine*, pp. 52–69 (February 2007)
23. K. Grenier, D. Dubuc, P.-E. Poleni, M. Kumemura, H. Toshiyoshi, T. Fujii, H. Fujita, Resonant based microwave biosensor for biological cells discrimination, in *BioWireleSS* (2010)
24. K. Grenier, D. Dubuc, M. Poupot, J.J. Fournié, Microwave signatures of alive B-lymphoma cells suspensions, in *BioWireleSS* (2011)
25. J. Lin, *Electromagnetic Fields in Biological Systems* (CRC Press, Boca Raton, 2012)
26. C. Gabriel, S. Gabriel, E. Corthout, The dielectric properties of biological tissue: I. Literature survey. Phys. Med. Biol. **41**, 2231 (1996)
27. M. Awayda, W. Driessche, S. Helman, Frequency-dependent capacitance of the apical membrane of frog skin: dielectric relaxation processes. Biophys. J. **76**(1), 219–232 (1999)
28. H. Schwan, K. Foster, RF-field Interactions with biological systems: electrical properties and biophysical mechanisms. Proc. IEEE **68**(1), 104–113 (1980)
29. M. Lazebnik, M. Converse, J. Booske, S. Hagness, Ultrawideband temperature-dependent dielectric properties of animal liver tissue in the microwave frequency range. Phys. Med. Biol. **51**(8), 1941–1955 (2006)
30. M. Schüßler, M. Puentes, D. Dubuc, K. Grenier, R. Jakoby, Simultaneous dielectric monitoring of micro fluidic channels at microwaves utilizing a metamaterial transmission line structure, in *Proceedings of 34th Annual International Conference of the Engineering in Medicine and Biology Society*, (San Diego, USA, August 2012)
31. M. Schüßler, M. Puentes, C. Mandel, R. Jakoby, Multi-resonant perturbation method for capacitive sensing with composite right-left handed transmission line, in *Proceedings of IEEE MTT-S International Microwave Symposium Digest*, (Anaheim, USA 2010)

Chapter 6
Analysis and Treatment of Organic Tissues

6.1 State of the Art in Analysis and Treatment of Organic Tissues

Microwave techniques to analyze dielectric properties of organic tissue, for the detection of abnormalities such as cancerous cells and its correspondent treatment is a hot topic for research. The current stage of the research is at understanding what information can be extracted and the design of appropriate devices and methods to carry it out. Although the applications are numerous, it is still very early to apply it in medical environments. The current methods for diagnosis are still based on computerized tomography (CT) scans, X-Rays or magnetic resonance imaging (MRI), and for treatment, surgery and chemotherapy are still the first choice for physicians [1]. These approaches still present many disadvantages and risks for the patient. In the case of CT and MRI, many patients experience discomfort and claustrophobia and need to be sedated, which can lead to complications. On the other hand, the use of these tests with children and pregnant women can be very complicated and, although the technology of these devices have advanced incredibly in the last years, false positives still occur. As for the treatment options, chemotherapy produces side effects in most cases, such as fatigue, nausea, vomiting, hair loss and others, and the success rate depends greatly on the type of cancer and stage. In the case of open surgery, some of the possible risks for the patient include anesthesia complications, bleeding, cardiovascular risks, infections and pulmonary risks. In general, there is no absolute cure for cancer. As a result there is a wide range of possibilities for new and innovative techniques in this field where microwaves and high frequency devices with their non-invasive principle can produce adequate methods for detection and treatment of different types of cancer. At the cellular level, Gascoyne has done extensive studies by using dielectrophoresis to separate healthy cells from cancer cells [2, 3]. There are even reports of the first steps to a completely integrated on-chip device that can use this technique to separate the cells by applying an AC field and obtain information not only about the cell morphology but properties of pools of cells without the use of markers or fluorescent agents [4, 5].

M. Puentes Vargas, *Planar Metamaterial Based Microwave Sensor Arrays for Biomedical Analysis and Treatment*, Springer Theses,
DOI: 10.1007/978-3-319-06041-5_6,

The study of dielectric properties of organic tissue can be the basis for developing improved techniques to detect these abnormalities and open the possibility for state of the art treatment mechanisms. As mentioned previously in Sect. 5.1, it has been proven that organic tissue presents a clearly identified dielectric behavior that is frequency dependent and it presents three major relaxation steps from low frequencies up to GHz. This was studied thoroughly by Schwan [6] and further explanations about the biophysical reasons for this steps were also analyzed [7, 8]. These initial studies were conducted mainly at the cellular level and not for the complete frequency spectrum. The studies of Gabriel [9] and Dean [10] offer a complete analysis of dielectric properties of most of the human tissues and organs at a very complete range of frequencies.

In the work presented in this chapter, it is desired to explore devices that can operate not at a cellular level but address organic tissues. By developing microwave devices and combining them with structures inspired by metamaterials, it can lead to a very cost-effective device that can localize with high precision an abnormality within the human body and maybe even provide treatment options for certain types of cancer [11, 12].

SRRs were used as building block for the devices that will be presented. They have been used in the past for sensing devices for applications such as pressure, temperature, humidity and concentration monitoring [13–15]. Another interesting application is the analysis of liquids and impurities through tracking resonant shifts and dips in the spectra of the SRR [16]. They have also been used in combination with mechanical structures for strain sensing [17] and to detect the progression of long-bone fracture healing in animals through strain [18]. Recently, research in the terahertz regime is having a wide impact in all areas and devices; sensors are not the exception and there is work done using SRRs in combination with thin-films for analysis of DNA [19].

In this chapter different microwave planar sensors based on SRRs will be presented with optimized designs to perform the sensing of the dielectric properties of different organic tissues and furthermore perform treatment of cancerous tissues with the same structures by means of thermal ablation treatments.

6.1.1 Frequency Characterization of Organic Tissues

The theory about biological material that was presented in Sect. 5.1 applies, of course, to organic tissues that are simply a macroscopic organization of biological cells. Nevertheless, a few words about the properties of the tissues itself will be given. Organic tissues are mainly composed of water molecules that tend to align themselves according to the applied EM field. Water percentage is different in different types of tissues and its influence can be observed in its permittivity, and therefore, the physical phenomena of EM energy dissipation is explained by the permittivity [20]. Tissues are considered heterogeneous lossy mediums for electromagnetic waves. At microwave frequencies the response of the tissues results in absorption of EM energy.

The coupling of the EM energy into the tissues is influenced by the geometry and composition of the exposed tissue as well as the frequency and configuration of the device used to deliver the energy.

At boundaries separating regions of different biological composition, the EM field is reflected or transmitted depending on the mediums intrinsic impedances. This can be represented by the following equations

$$R = \frac{(\eta_2 - \eta_1)}{(\eta_2 + \eta_1)} \tag{6.1}$$

$$T = \frac{(2\eta_2)}{(\eta_2 + \eta_1)}, \tag{6.2}$$

where R is the reflection coefficient, T is the transmission coefficient and η was defined in Eq. (2.14). Both parameters are related to each other by $T = 1 + R$. For similar tissues, very little reflection will take place and most of the energy will be transmitted. As the transmitted field propagates into the tissue a reduction in the power density of the field will take place and this RF energy will be deposited in the tissue. The penetration depth is about five times higher when the tissues have a low water content like bone or fat in comparison to a high water content tissue such as most solid organs like heart, lungs, liver, spleen or kidneys. On the other hand, the transmission coefficient doubles for air-tissue interfaces when the tissue has a low water content compared to high water content tissue. Moreover, the transmission coefficient is larger for tissue-tissue interfaces than air-tissue interfaces, whereas the reflection coefficient shows the opposite trend. In tissues with several layers all having different permittivities, the coupling behavior is very complex. Multiple reflections can occur in all the tissue interfaces and the transmission coefficient will combine all transmitted and reflected waves forming standing waves in each tissue layer. At the peaks of the standing waves a higher coupling can take place on the specific layer and this phenomenon becomes critical when the thickness of the layer is greater than the skin depth for that tissue layer, approximately $1/2$ of the guided wavelength λ_g or longer resulting in a resonant behavior [21].

In summary, when electromagnetic fields are coupled into tissues a biological response takes place that must be quantified and correlated with the observed effect. The most commonly used parameters for this is the specific absorbtion rate (SAR). SAR is defined as the power deposited in a unit mass of tissue, and can represent the biological activity of the tissue with respect to EM energy exposure. It is represented by [22]

$$SAR = \frac{\sigma(r) | E(r) |^2}{2\rho(r)}, \tag{6.3}$$

where $\sigma(r)$ is the MUT electrical conductivity, $E(r)$ is the RMS electric field and $\rho(r)$ is the MUT density. SAR is given in (W/kg) and can be used for both near and far field scenarios. It is useful, because it allows for the extrapolation of experimental results to different configurations and the analysis of the relation between different experimental models and subjects.

In medical research and practice the dielectric properties of tissues are without doubt of high importance. As the dielectric properties of healthy and malignant tissues differ considerably since their water content is quite different, the phenomenons presented in this section can be used to diagnose the malignant tissues surrounding healthy tissue and in some cases perform treatment as well.

6.2 1-Dimensional Sensor Array for the Analysis of Organic Tissues

6.2.1 Sensor Design and Fabrication

The main goal is to design planar microwave sensor arrays that are able to interact with organic tissue in a non-invasive way and extract its dielectric image with high resolution and accuracy. In Fig. 6.1 is an example of a possible application for these sensors. The sensor comes into direct contact with the tissue to be analyzed through a minimal invasive procedure and extracts a relative dielectric image. Since the goal is to use these structures as part of a minimal invasive tool, their size and miniaturization possibility is an important parameter in the design. The possibility to have a dielectric image with several pixels of resolution is also desired and was another reason for choosing metamaterial structures in the design. The initial design to be considered was a 1D array of SRRs excited by a microstrip line as shown in Fig. 6.2.

The reasons for choosing SRRs are their strong response to an electromagnetic field, their easy miniaturization and high Q performance. The equivalent circuit of a microstrip line loaded with a single SRR [24] for the quasi-static case was shown in Fig. 2.16. The SRR is magnetically coupled to the transmission line with the coupling factor M. L_S and C_S correspond to the total SRR inductance and capacitance, respectively. The sensor array is made of several SRR particles that can produce a spatially resolved dielectric image of the MUT. The image will have as many pixels as SRRs. For this purpose, it is necessary that the resonant frequency of each pixel can be controlled independently to be able to multiplex each sensitive region and obtain a separated analysis from the section of MUT surrounding each SRR. The main parameter to be considered in the design of the prototype sensors is to avoid any coupling or influence between the rings. With this achievement, a separate analysis of the perturbation at each sensitive region is possible and the sensors have the possibility to extract dielectric properties independently on each sensitive area according to the amount of unit cells in the prototype. To obtain this independent behavior, the geometry of the SRRs has to be altered. To simplify the process, it was decided to change only one parameter and the selected one was the width of the SRR particle. This was shown in Sect. 2.3.1.

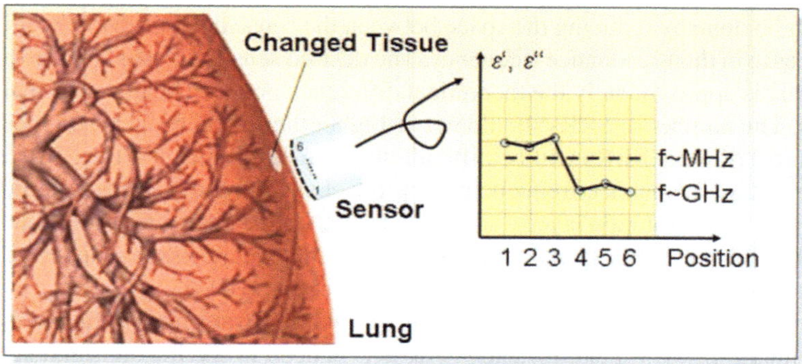

Fig. 6.1 Planar microwave sensor array for analysis of organic tissue and the extracted dielectric image with 6 pixels of resolution

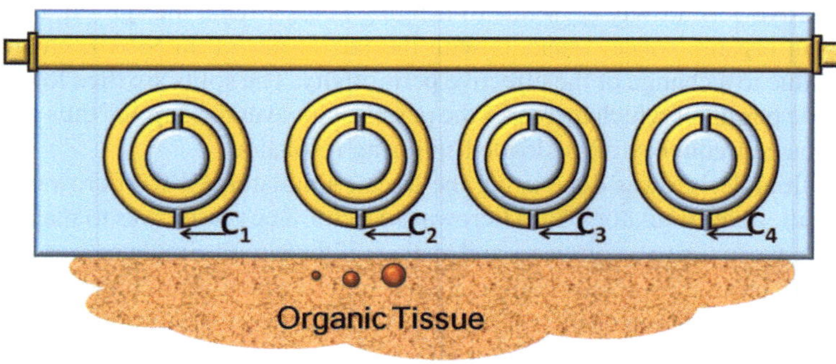

Fig. 6.2 Overview of the 1D sensor array with 4 pixels made of microstrip excited SRRs [23]

Table 6.1 Design parameters of the 1D sensor array for analysis of organic tissue

Parameter	Desired value
Center operation frequency	2 GHz
Sensitivity $(\Delta\varepsilon/\varepsilon)_{min}$	10 %
Dimensions (sensor)	Length = 7 cm; Width = 5 cm; Height = 3 mm
Number of SRRs	4
Isolation layer	Teflon

The design parameters for the 1D sensor array are given in Table 6.1.

The reason for the chosen values of the design parameters will be addressed. The center operation frequency was selected to be around 1 GHz for the initial prototype to facilitate the fabrication and measurement with bulky tissues but keeping in mind the miniaturization of the final tool which should not be a problem with the SRRs. To reduce the size, it would be necessary to decrease the electrical size of the SRRs and this can be done by reducing the space between the rings and their width resulting in an increase in their resonance frequency. The desired sensitivity was selected at 10 % since that is approximately the theoretical difference between healthy and malignant tissues. The number of SRRs should not be higher than 4 for initial prototypes, since the control of more SRRs will be difficult during measurements if the SRRs are too similar in their resonance frequencies or if the change in the loading material is too high. Finally, an isolation layer was clearly needed in the design to protect the sensor from the high conductivity and inherent losses of the organic tissues and teflon seems to be a simple material and easy to obtain for this purpose.

As it was already explained, an SRR is excited by a transmission line with a CW signal and a resonant peak for each SRR is produced in the transmission spectrum due to coupling. The SRRs are positioned such that the outer splits are opposite to the line. Those splits can interact with a pertuber for which the permittivity is to be measured. Roughly speaking, the split of an SRR can be described as a capacitor with effective permittivity $\epsilon_{r,eff}$ due to fringing fields around the split. If a pertuber, such as an organic tissue, interacts with the outer split of an SRR it changes the capacity due to a change of the effective permittivity. The splits are then loaded with a dielectric perturber which increases the permittivity over the gap and thus decreases the resonance frequency of the loaded split ring resonator.

The initial design was done using the circuit simulator AWR Microwave Office and the full wave simulator CST Microwave Studio. The reason was to make a comparison between the two softwares and analyze if the circuit simulator was sufficient in terms of accuracy since its computational time is much shorter and the aim was to make several different designs. For the AWR Microwave office a careful selection of elements that best represent the actual structure was done. An example is shown in Fig. 6.3. For the coupling between the feeding line and the SRR an edge couple microstrip line element was chosen and the length, width and gap distance can be controlled. Critical for the layout was an appropriate selection of corner element to represent the square shape of the SRRs. The microstrip bend with optimally mitered arbitrary angle was selected and it is believed that the slight discrepancies with the full wave simulations come from this particular element. In Fig. 6.4, a comparison between CST Microwave Studio and AWR simulations for the 2-Double-SRR (sensor with two unit cells made of double rings) is shown and although a slight shift in frequency is present, a good agreement is achieved. From this result it can be concluded that a circuit simulator is adequate and can replace a full wave simulation; even the coupling effects between multi-SRRs cells are analyzed precisely.

To proof the concept, four different sensor prototypes were designed. An SRR sensor based on single rings for two and four unit cells and an SRR sensor based on double rings (an inner and an outer ring) also for two and four unit cells were developed. The reason for this was to analyze which geometry is more adequate in terms of performance and simplicity to extract the relative dielectric values. The final prototypes where constructed as shown in Figs. 6.5 and 6.6. They were built on Rogers RT/Duroid 5880 with dielectric constant of $\varepsilon_r = 2.2$ and a thickness of $h = 1.57$ mm. The structure is a microstrip design and the sensitive area corresponds to the gap of the outer rings in each unit cell. In Fig. 6.5, a picture of the prototypes of

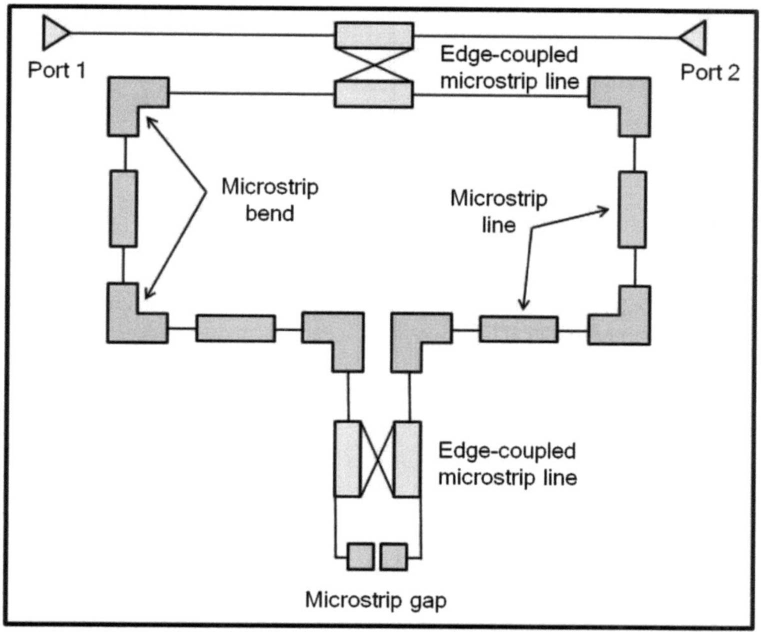

Fig. 6.3 Schematic design of a 1-Single-SRR. The software used is AWR Microwave office

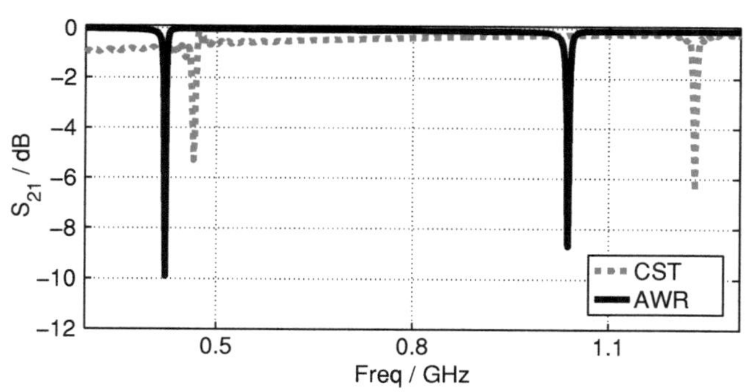

Fig. 6.4 Comparison between circuit simulator and full wave simulator for the case of the 2-Double-SRR [23]

2 pixels is shown and the dimensions of the SRR are highlighted. On some of the SRR a tuning capacitor was introduced to control the frequency and separate the resonant peaks between SRRs. The prototypes with four unit cells are shown in Fig. 6.6. To compare the single ring SRR and the double ring SRR a sensitivity analysis was carried out. The change in the resonant frequency in dependance of the change in the dielectric constant was calculated. The results from the sensitivity analysis are shown in Table 6.2. Although the double ring design has a higher figure of merit, from the

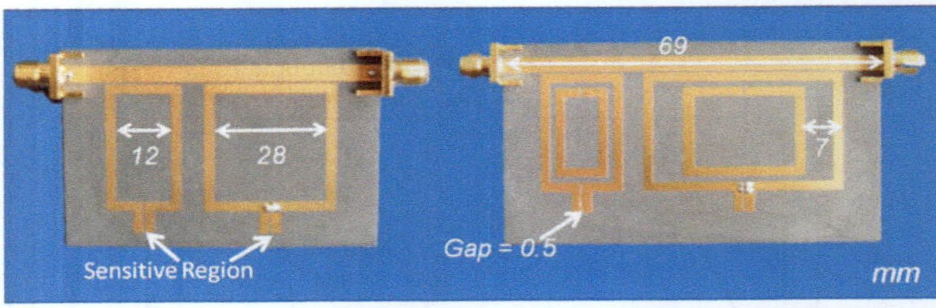

Fig. 6.5 1D sensor prototypes with two unit cells. On the *left figure*, the SRR are made of a single ring and on the *right figure* they are made of two concentric rings (the dimensions are highlighted in mm as well as the sensitive area) [23]

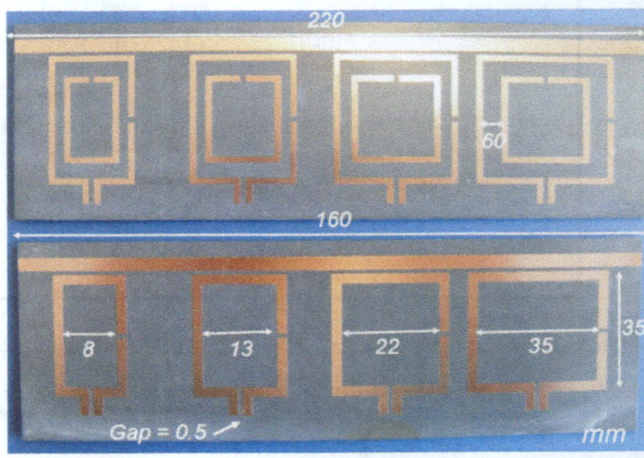

Fig. 6.6 1D sensor prototypes with four unit cells. On the *upper figure*, the SRR are made of two concentric rings and on the *lower figure* they are made of a single ring (the dimensions are highlighted in mm)

Table 6.2 Sensitivity analysis results of the 1D sensor array prototypes

Prototype	Sensitivity = $\Delta f/\Delta\varepsilon$
2-Single-SRR	4.8
2-Double-SRR	5.2
4-Single-SRR	6.0
4-Double-SRR	6.6

table, it can be inferred that all the options are suited for the desired application since the difference in their sensitivity is quite small. A maximum of 1.8 % can be seen between the 2-single-SRR and the 4-Double-SRR. Therefore, for most of the devices that will be presented in the following sections, the single ring option will be used due to its simplicity in the design process and this is very important for the 2D array where many resonators have to be controlled.

Finally, in Fig. 6.7, a comparison between simulation and measurement for the prototype with 4 pixels, each made with the double ring geometry (4-Double-SRR sensor), is depicted and a good agreement was achieved. The slight difference in the frequency is mainly due to the approximations made in the simulation process when using the circuit simulator and to fabrication tolerances.

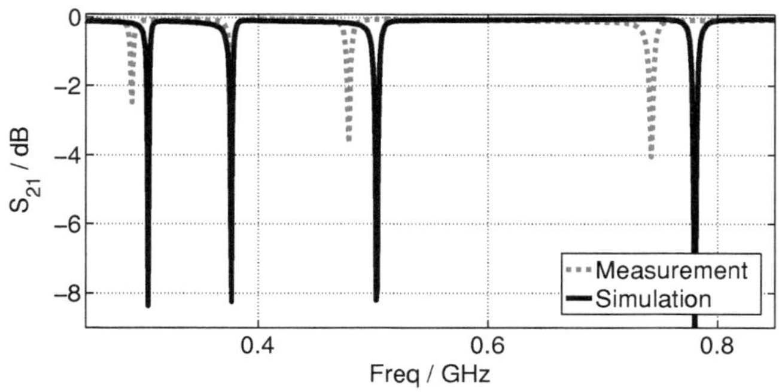

Fig. 6.7 Comparison between measurement and simulations for the 4-Double-SRR sensor. A good agreement is achieved despite the slight shift in frequency [23]

6.2.2 Capacitive Extraction Model

Two models were developed for the extraction of the dielectric image. The first is a capacitance extraction model and the second a permittivity extraction model. For the capacitance extraction model, the theory of resonators was used by only taking into account the first order resonances and by modeling both structures, the single and double ring case, as an equivalent resonant circuit as shown in Fig. 2.16a with resonance frequency (2.61). Since we are only interested in the influence of the tissue on the capacitance and how it can be tracked through the resonant frequency of the SRR, we consider no losses here. To take into account the gap capacitance C_{gap} and the perturber capacitance $C_{\text{perturber}}$, Eq. (2.61) is rewritten as

$$\omega_s = \frac{1}{\sqrt{L_s}(C_{\text{gap}} + C_{\text{perturber}})}. \tag{6.4}$$

Solving for $C_{\text{perturber}}$ yields

$$C_{\text{perturber,n}} = \frac{1}{L_s \omega_s^2} - C_{\text{gap}}. \tag{6.5}$$

To obtain the unknown values of L_s and C_{gap}, a parameter sweep of different perturber capacitances is made and the corresponding resonance frequencies are analyzed. Furthermore, model fitting is applied to find the effective inductance value, which leads to a set of parameters for each ring. From this model, we can derive the perturbing capacitance value by observing a certain shift of the resonance frequency. In Table 6.3 the corresponding values are given for each of the rings of the 1D prototype sensors. An example of the curve fitted model for the first ring of the 2-Single-SRR structure is shown in Fig. 6.8 and this was realized for all rings of all prototype structures.

Table 6.3 Equivalent circuit values for all 1D prototype sensors [23]

Prototype	n = Ring	L_{ring} (nH)	C_{gap} (pF)
2-Single-SRR	First	39.90	2.99
	Second	28.70	0.64
2-Double-SRR	First	50.00	2.85
	Second	32.8	0.71
4-Single-SRR	First	37.00	4.32
	Second	31.62	3.19
	Third	27.00	1.77
	Fourth	23.80	0.60
4-Double-SRR	First	52.65	5.20
	Second	47.50	3.77
	Third	43.50	2.30
	Fourth	35.50	1.16

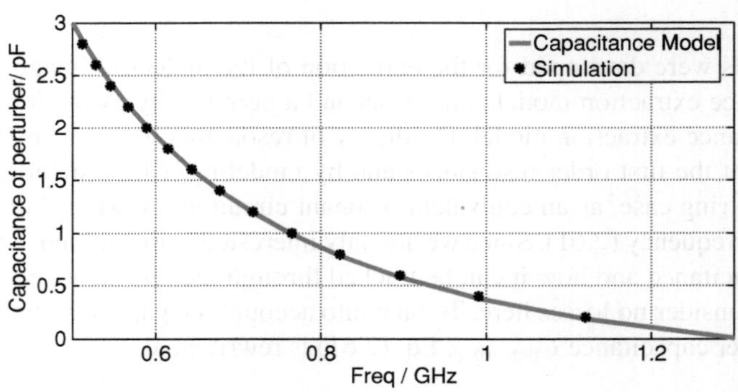

Fig. 6.8 Capacitance model for the first pixel of the 2-Single-SRR prototype. A good agreement between the developed model and the simulations is achieved [23]

6.2.3 Permittivity Extraction Model

For the extraction of the dielectric properties, a model was constructed based on the simulations from the software CST Microwave Studio. Each of the four prototype sensors was simulated with a perturber on each of the sensitive areas. The perturbers have a cubic shape and they are made of teflon with different dielectric constants, from $\varepsilon_r = 2.2$ to $\varepsilon_r = 25$, and a size of approximately two times the gap of the SRRs in every direction, that is, $2 \times 3 \times 2$ mm^3. In total six different perturbers where used and they were placed on each of the SRRs of each prototype and the change in the resonant frequency of the specific SRR was monitored. This information was plotted and linearized using a simple fitting tool to create the permittivity model for each SRR of each prototype independently. An example is shown for the 2-Double-SRR structure in Fig. 6.9.

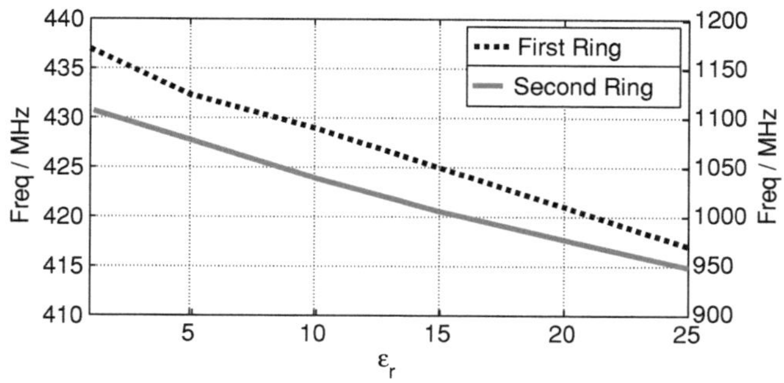

Fig. 6.9 Permittivity model for the 2-Double-SRR structure. The model is shown for both SRRs [23]

The presented model is valid only for the specific size of the perturber and relative position to the sensitive area of the prototype sensor. In case that any of these parameters are altered, the model has to be recalculated. This can be highly complex as the amount of pixels increases but since we are only interested in relative dielectric changes, for a large number of pixels, the capacitance model is sufficient. This is due to the fact that for monitoring relative dielectric changes to detect the location of altered tissue with permittivity different to normal tissue, the absolute permittivity value is not necessary and by extracting the changes in the capacitance, these abnormalities can be located.

6.2.4 Measurements Results

To prove the concept, different measurements were performed on all of the prototype sensors. Thinking ahead for the 2D array, the first test is to analyze the behavior of a each single SRR within the array to proof that it behaves independently from its neighboring SRRs and shifts its resonant frequency when perturbed. In Fig. 6.10 it is shown how each ring of the 2-Double-SRR sensor responds when it is perturbed with the same cubic dielectric bricks used previously for the permittivity model. These perturbers, as mentioned, range from values of $\varepsilon_r = 2.2$–25 and will be used in all the dielectric loading measurements. Since the resonant frequency range of each ring is different, the plot has been normalized to the unloaded resonance value of the particular SRR.

Figure 6.11 shows a measurement result made with the 4-Single-SRR. The dielectric perturbers were placed on the SRR one and four of the structure with values of $\varepsilon_r = 2.2$ and 10.6, respectively. In the inset of the figure, the changes on the resonant frequency of the perturbed rings are depicted while the other remain undisturbed. The extracted values of the profile given in Table 6.4 were obtained using the permittivity model. The results are within the tolerance limits established in the design

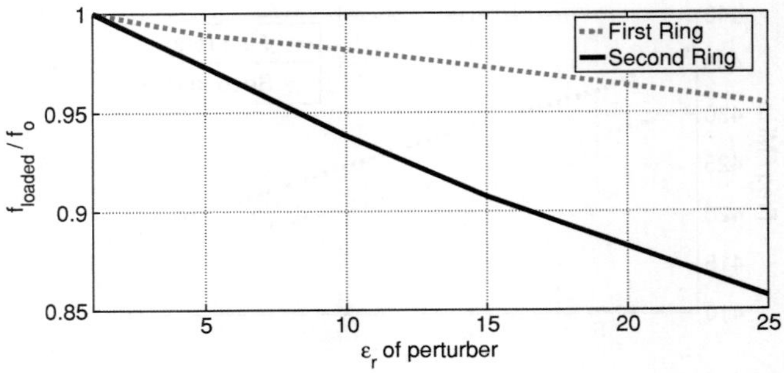

Fig. 6.10 Resonance frequency behavior of the SRR for the 2-Double-SRR structure. The frequency is normalized to the unloaded resonance frequency of each SRR to show the case for the first and the second pixel. The shift in the individual resonant frequencies is clear as the ε of the perturber changes

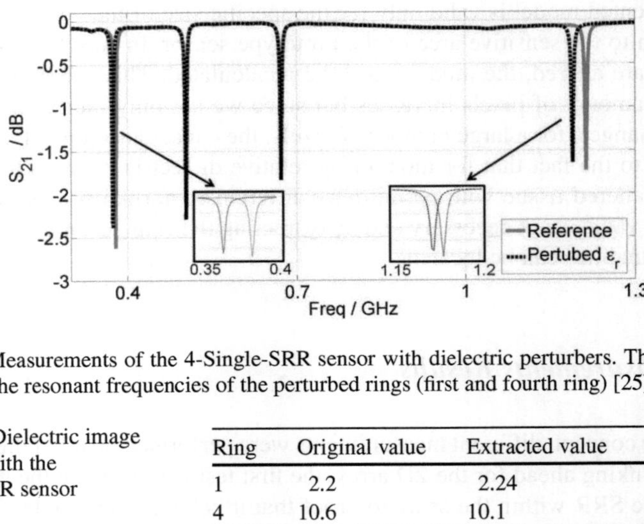

Fig. 6.11 Measurements of the 4-Single-SRR sensor with dielectric perturbers. The inset shows the shift of the resonant frequencies of the perturbed rings (first and fourth ring) [25]

Table 6.4 Dielectric image extraction with the 4-Single-SRR sensor

Ring	Original value	Extracted value	Relative error
1	2.2	2.24	0.018
4	10.6	10.1	0.049

phase. It should be mentioned that the placing of the MUT is very sensitive, any slight movement of the MUT results in an influence on the resonance frequency. For this reason a special setup where the precise placement of the MUT can be done is necessary and will be addressed in the following sections.

After the structures and the different models were tested with the dielectric perturbers, additional measurements were performed with animal lung tissue. Measurement results for the 4-double-SRR sensor loaded with lung tissue are depicted in Fig. 6.12. From the resonant peak it is clear that the lung has very high losses in comparison to the dielectric perturbers, this can be seen by comparing the Q factor and extracting tan δ from the lung. The Q factor changes from 157 from the unloaded

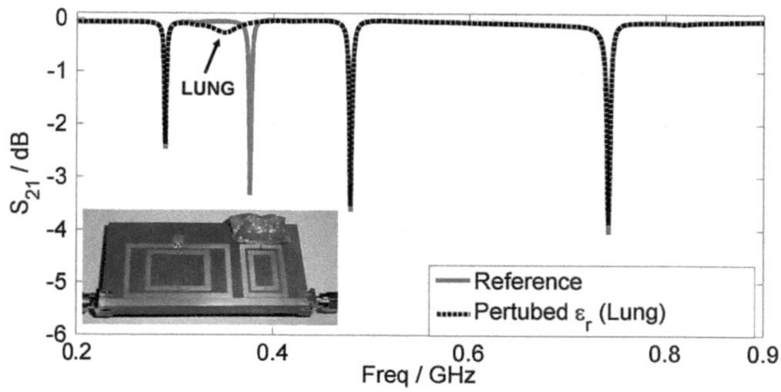

Fig. 6.12 Measurements of lung tissue with the 4-double-SRR sensor [23]

sensor to 10 when the lung is measured and its extracted $\tan\delta = 0.1$. Nevertheless, the resonant frequency can be clearly identified and the extraction can be done. After applying the permittivity model, the permittivity of the lung tissue was calculated to be $\varepsilon_r = 51$ which for the frequency of operation of the sensor is in good agreement according to the literature [9]. This was possible since the size of the lung tissue is small in comparison with the size of the sensor and it is position in a way that it does not come into contact with the SRR. This careful positioning can be very difficult since the tissue have a gelatine-like composition. On the other hand, this also affects the repeatability of the measurements since positioning the tissue exactly in the same location as in a previous measurement is impossible. Additionally, if the tissue directly touches the metallization of the structure the signal will be lost due to the high conductivity of the tissue. To solve this difficulty an easier approach is to slightly press the tissue on top of the complete structure, therefore, covering it completely and for this purpose an isolation layer will be included in the following designs as done for the microfluidic sensor of Chap. 5.

6.3 2-Dimensional Array for the Analysis of Organic Tissue

The satisfactory results from the 1D array, where relative dielectric monitoring with a sensitivity of 10 % was successfully carried out, initiated the design of a 2D array that could make the analysis of the dielectric properties of the tissue in two dimensions and straightforwardly provide information on the location of abnormalities within the tissue. The design process was similar to the 1D array and an overview of the 2D sensor array can be seen in Fig. 6.13. This further development includes 12 SRRs, the rings are detuned, decoupled from each other and have their own resonant peak which is possible again by changing the width between pixels as seen in Fig. 2.12. This array will provide a spatial distribution in two dimensions were the detection of local relative changes in the dielectric properties of the tissue can also be made with

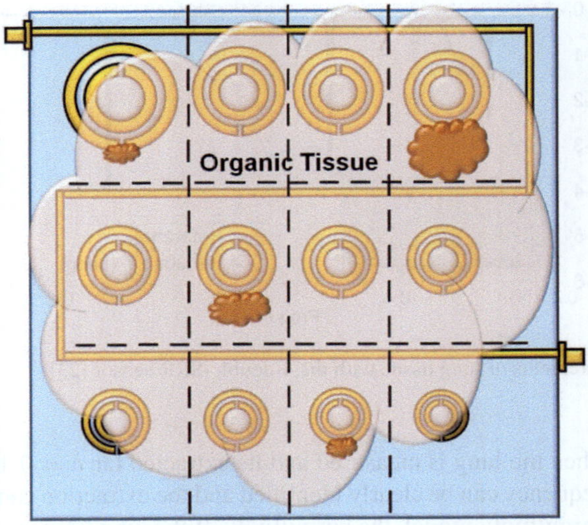

Fig. 6.13 Overview of the 2D sensor array. It has 12 pixels in a 4 × 3 configuration [26]

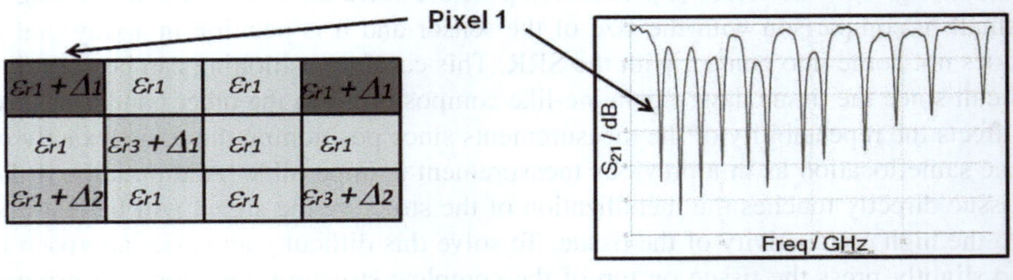

Fig. 6.14 Dielectric image extraction. Each peak in the scattering parameters corresponds to one SRR or pixel that will extract one relative dielectric value to obtain the complete dielectric image [26]

the same procedure of tracking the resonant frequency as shown in Fig. 6.14. Here, each SRR has an independent resonance frequency that corresponds to a pixel of the dielectric image. A cross section of 1 pixel is shown in Fig. 6.15; here an isolation layer is again included between the SRR and the tissue to prevent the degradation of the resonant peaks due to the very high conductivity inherent to the tissue.

The prototype is shown in Fig. 6.16 with its corresponding dimensions. By comparing the size of the first ring of each structure, for the 1D array it is 5×3 cm^2 and for the 2D array is 3×4 mm^2, a considerable reduction in size is achieved. As a result the operating frequency is of course higher for the 2D array. This is an interesting feature since the final goal of these devices is to be included in minimal invasive tools and for this purpose the miniaturization feature is important. Fortunately, the SRRs are very flexible and provide this possibility.

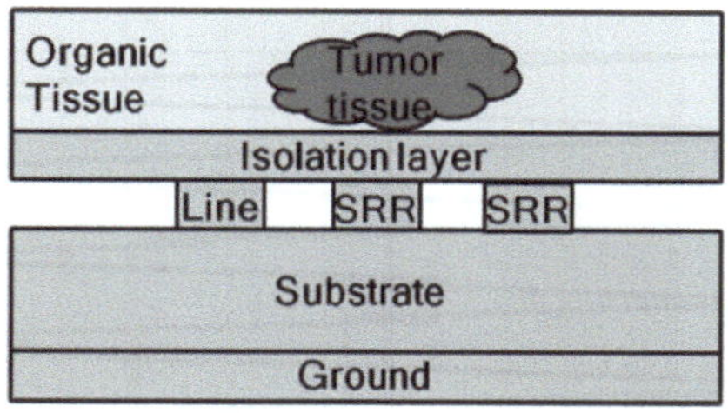

Fig. 6.15 Cross-section of 1 pixel. The isolation layer is located between the sensor and the organic tissue to prevent the degradation of the resonant peaks due to the inherent losses of the tissue [25]

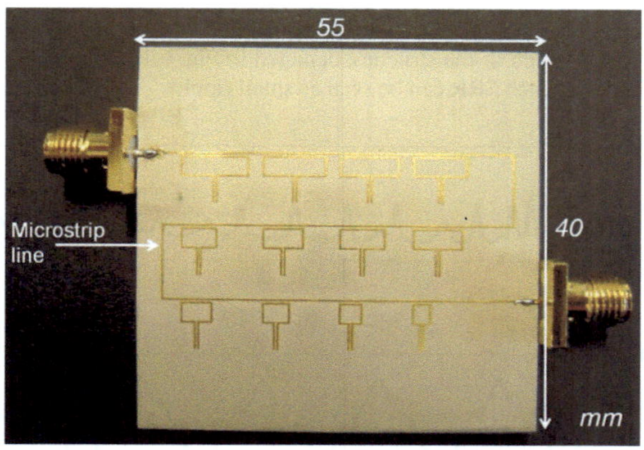

Fig. 6.16 Prototype of the 2D sensor array. It was made on substrate RT/duroid 6010 with $\varepsilon = 10.2$ and thickness of 0.254 mm [25]

In Fig. 6.17 the dispersion diagram of the structure is shown. It behaves as a transmission line and the resonances of each SRR can be seen as small ripples along the line. In Fig. 6.18 the propagation parameters of the structure are depicted. From the plots of α and β it can be once more corroborated that it behaves as a transmission line. On the other hand, the resonances of the SRRs can be seen at the same frequencies that are given in the scattering parameters of the structure as will be shown in Fig. 6.22.

6.3.1 Measurement Setup

Although tests for the 1D array structure were successful, it was noticed that the position of the sample of tissue was critical to obtain good results and the repeatability of the measurements was compromised by changes in the placement of the same sample. For this purpose a measurement setup was developed. The design centers on

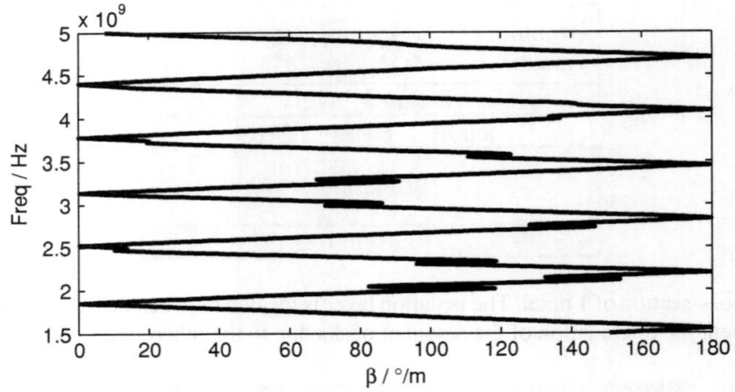

Fig. 6.17 Dispersion diagram of the structure depicted in Fig. 6.16. It behaves as a transmission line and the resonances of each SRR can be seen as small ripples

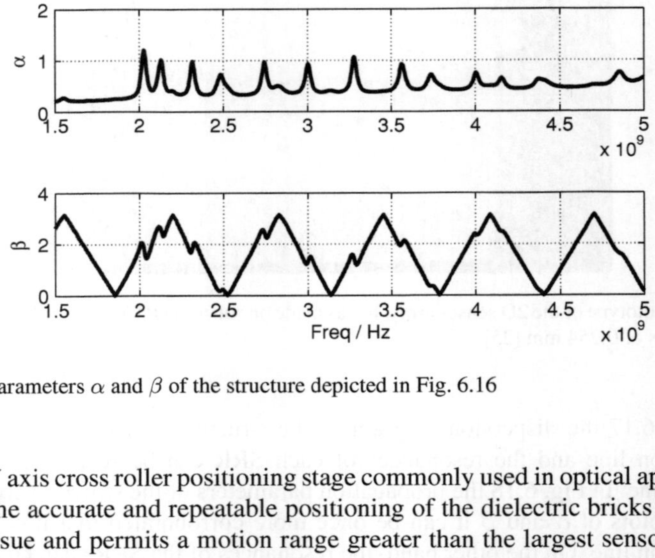

Fig. 6.18 Parameters α and β of the structure depicted in Fig. 6.16

an X and Y axis cross roller positioning stage commonly used in optical applications. It allows the accurate and repeatable positioning of the dielectric bricks and/or test organic tissue and permits a motion range greater than the largest sensor that is to be tested with the measurement setup. The positioning stage achieves this by being controlled by micrometer heads in the X and Y axis of motion, which have accuracy and repeatability of less than 10 μm. When conducting measurements, the cross roller positioning stage is clamped down firmly to a base.

For the measurement setup to have the flexibility of holding different sensors as large as $60 \times 45\ \text{mm}^2$ to a minimum of $53 \times 25\ \text{mm}^2$, a two tier approach was used in the design. The first tier consists of two plastic clamp/hold pieces, which purpose is to support the sensor and provide two fixed edges along the sides of the sensor. The second tier consists of a metal screw capped with a polyacetal part that is inserted from the distal end of the movable clamp relative to the sensor. The purpose is to firmly clamp the sensor, holding it rigidly throughout multiple measurement tests with no metal contacting it at all times.

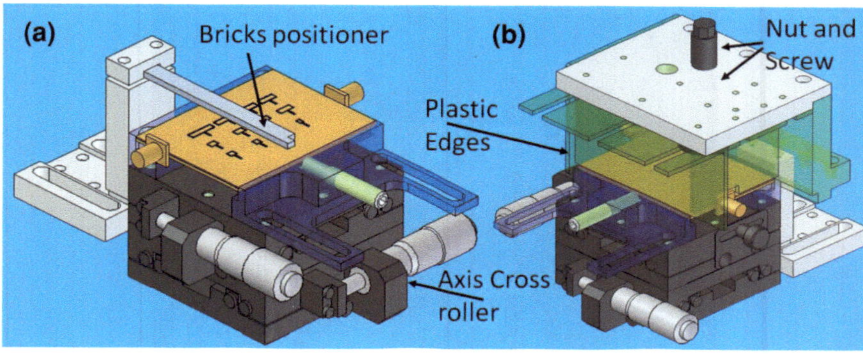

Fig. 6.19 **a** Measurement setup to position dielectric bricks on the prototype sensor **b** Organic tissue measurement setup. The housing for the organic tissue is shown along with the screw to press it and eliminate air gaps [25]

In Fig. 6.19a is shown a schematic of the complete setup to position the dielectric bricks. It includes a guide made out of sheet metal that creates a stationary reference edge. The guide is maintained stationary by clamping it down to an aluminum base.

In Fig. 6.19b a schematic of the organic tissue measurement setup is shown. The sheet metal guide is removed, yet its base remains, and one of two (non-identical) plastic master edge parts is attached to the base and extend the master edge for 35 mm above the stationary clamp. It allows the correct alignment of the organic tissue samples. On the top, an aluminum base is attached to apply pressure on the organic tissue to avoid any air gaps. The aluminum base is sufficiently high in order to prevent electromagnetic disturbances during testing of organic tissue. Similarly, two other plastic parts are attached around the other two edges of the sensor that extend 35 mm. This allows organic tissue to be tested at thicknesses up to 30 mm. Because the sensors vary in size and positioning of the coaxial cables, multiple drill and tap holes have been inserted into the metal aluminum base to allow the adjustment of two remaining plastic side walls.

6.3.2 Measurements with the 2D Sensor Array

The measurements for the 2D sensor array prototype were done by using the proposed measurement setup as shown in Fig. 6.20. The first test is to observe if the rings are behaving independently like for the 1D array. For this purpose a perturber of $\varepsilon_r = 10$ was located on each ring at a time. Figure 6.21 shows how only the perturbed ring is affected while the others remain undisturbed. Although all the rings were perturbed with the same dielectric perturber, it is interesting to see that the change of each ring is different, they all respond to a constant change by a unique change in the resonant frequency. In the analysis of the results, this characteristic has to be taken

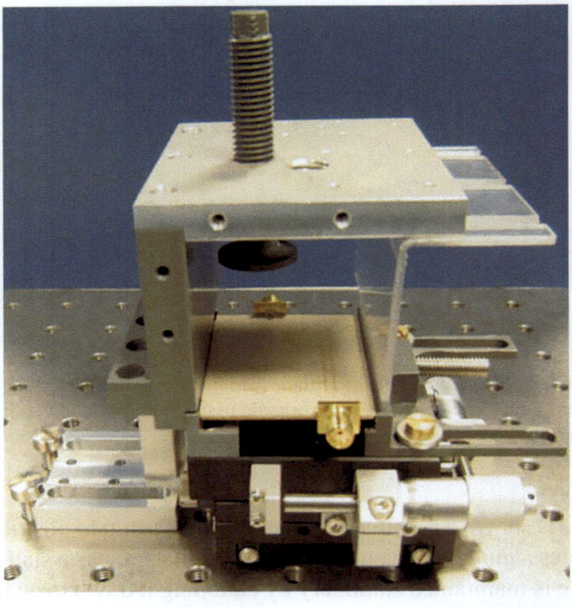

Fig. 6.20 Prototype of the 2D sensor array mounted on the measurement setup for organic tissue [25]

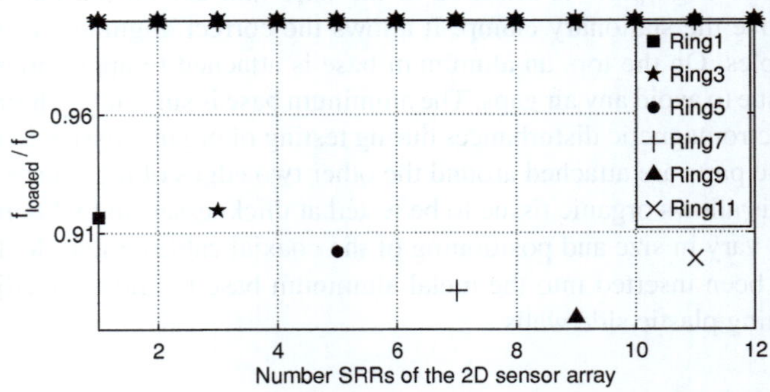

Fig. 6.21 Independent behavior of the SRR for the 2D sensor array. Each ring was perturbed separately with a dielectric brick. In all cases, only the frequency of the altered pixel changed as the others remain undisturbed [25]

into account and a normalization is required in the extraction of information from the organic tissue.

To prove the concept, different profiles were created with the dielectric bricks. A typical scenario is shown in Fig. 6.22 where a dielectric load of $\varepsilon_r = 2.2$ was placed on the third and tenth ring and an additional load of $\varepsilon_r = 10.2$ was located on the twelfth ring. The shift of the resonance frequencies of the three rings can be seen and the other frequencies of the undisturbed rings remain steady. The extraction methods

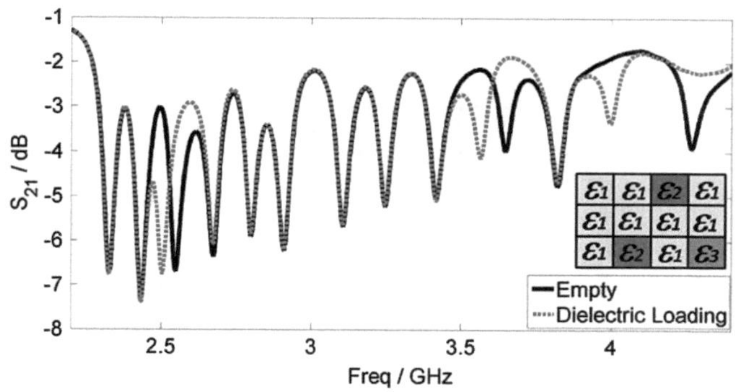

Fig. 6.22 Measurements of dielectric perturbers loading with the 2D sensor array. The third, tenth and twelfth pixel were loaded. The frequencies of the other pixels are not changed [26]

Table 6.5 Dielectric image extraction with the 2D sensor array

Ring	Original value	Extracted value	Relative error
3	2.2	3.7	0.68
10	2.2	4.5	1.04
12	10.2	18.7	0.83

presented previously were applied and the dielectric values of the perturbers were extracted and are shown in Table 6.5. The relative error is also shown in the table and although it has a high value, it is possible to extract the profile and it should be emphasized that the goal of this sensors is to extract the relative change based on an initial calibration measurement and not an absolute value.

Another test was conducted with animal lung tissue. For these measurements, the already mentioned dielectric isolation sheet of $\varepsilon_r = 2.2$ and thickness of 254 µm is placed between the array and the organic tissue to prevent the degradation of the resonant peaks due to the high conductivity and losses from the tissue. The results are shown in Fig. 6.23. The first curve represents a sample of lung placed on top of the isolation sheet that covers practically the whole area of the sensor. The second curve represents the same lung tissue but with an additional dielectric perturber of $\varepsilon_r = 2.2$ placed under the tissue around the second and third pixel. The shift of the frequency peak can be clearly identified, and therefore, it is proven that a monitoring of local changes in the tissue samples is possible. The sensors presented here operate with 1 mW of power. This represents no danger to the tissue of being heated which translates in a non-invasive measurement.

6.3.3 *Phantoms of Human Tissue*

After making measurements with dielectric loading and animal lung tissue, it was desired to test if the prototypes will give the same results with human tissue. The

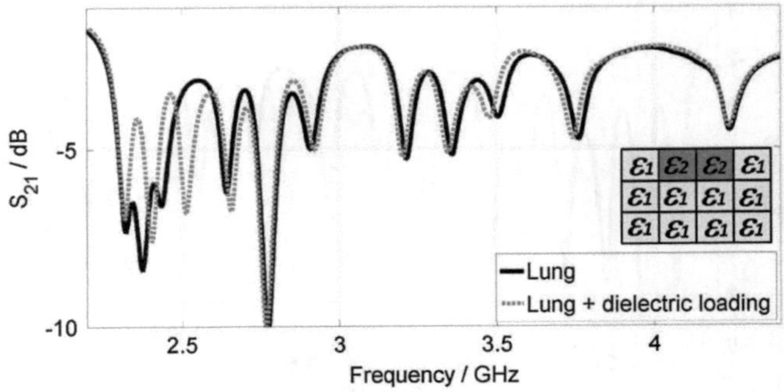

Fig. 6.23 Measurements of animal lung tissue with the 2D sensor array. Dielectric loading was additionally included around the second and third pixel. The isolation layer clearly fulfilled its purpose and there is no significant degradation of the resonant peaks [26]

differences between animal tissue and human tissue, and furthermore to in-vivo tissue are considerable. The lifespan of ex-vivo human tissues is short and it needs extreme conditions to be used, preserved and maintained. For these reasons, the need to make physical phantoms for conducting experiments is evident. They should have the same dielectric characteristics and dispersive behavior as real tissue. The physical phantoms exhibit similar dielectric properties as biological tissues when exposed to EM energy, but usually these are made using non-biological materials. The most important features they should have are mechanical stability, non-toxicity, simple fabrication, wideband properties, long life and low cost. The recipes for breast tissue phantoms presented in [27] were selected as they comply with the desired features. Nevertheless further investigations were made on how to adapt it for other tissue types. Further information on the fabrication process and recipes can be found in Appendix A.1.

The phantoms are based on glycerin and by changing the ratio of water, salt, agar and poly-ethylene powder they can mimic any type of tissue. Agar works to add mechanical strength to the polymer matrix of the phantom. As water content is the major contributor to the dielectric properties of the tissues, glycerin phantoms can reflect properties of high water content and low water content tissues by changing the percentage of salt and water. Water reflects the permittivity and salt (NaCl) reflects the conductivity. Polyethylene powder (PEP) is added for improving the quality. The glycerin phantoms are chosen to be used in this study as they exhibit the moisturizing effect, which leads to less evaporation of the water contents retaining the properties for a longer time. They can be applicable for up to 6 months if stored in refrigerator with sealing film wrapping. From the studies presented in [28], losses of phantom materials like agar, water and glycerin are larger from frequencies above 3 GHz and there is no need to add salt to the recipe. The phantoms are shown in Fig. 6.24.

To be able to use the phantoms and mimic different tissues as desired, it was important to characterize the phantoms after changing the percentage of the different ingredients and check its dielectric properties. The measurements of the dielectric

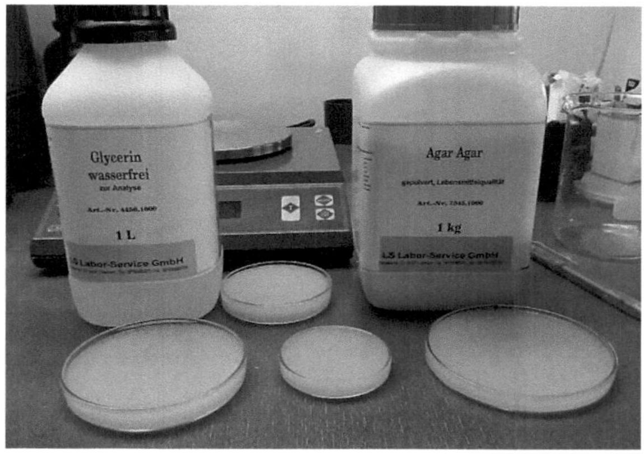

Fig. 6.24 Developed phantoms of organic tissue

properties of the phantom are done using 85070E dielectric probe kit and a vector network analyzer, which measures the microwave energy response. The probe transmits a signal into the MUT, with a frequency sweep up to 20 GHz. It has an easy to use software with graphical user interface and simple calibration procedure. The calibration process is guided by the GUI with an air/short/water standard. The measurements are done from 1–20 GHz with 1600 equally spaced frequency points. The tests were performed by changing the percentage of polyethylene powder (PEP) from 10 to 18 % and the results are shown in Fig. 6.25. It is clear that the dielectric constant of the phantoms can be altered by changing this ingredient. Similar results were obtained by changing the percentage of the agar powder as seen in [29]. Since fat has low dielectric constant values, it can be mimicked using lower percentage of PEP or agar. On the other hand, muscle like tissues have a high dielectric constant and can be prepared using higher percentage of PEP or agar. At 5 GHz the dielectric constant of the phantom with 10 % PEP is approximately 25 and with 18 % of PEP it has a dielectric constant of approximately 33. In Fig. 6.26 is shown a measurement result with the 2D sensor array of a phantom that behaves as fibro-glandular tissue in solid black line and the dashed grey line represents a measurement of the same fibro-glandular phantom with malignant pieces inserted around the seventh, eighth and ninth unit cell. In both cases the phantom covers almost the complete sensor and only the resonances of the pixels that interact with the malignant tissue show a significant change proving once more the performance of the presented sensor structure. Nevertheless, it was not possible to establish how close from the gap has to be located the perturber in order to be detected. The exact placement of the malignant tissue within the phantom is not easily defined. This is mainly because in the fabrication of the malignant phantoms it is used alginate powder and it solidifies within 10 s making it very difficult to make defined shapes. Since the malignant phantoms do not have a defined shape and are so hard to insert in the normal phantom its exact positioning in respect to the SRRs is no systematically defined.

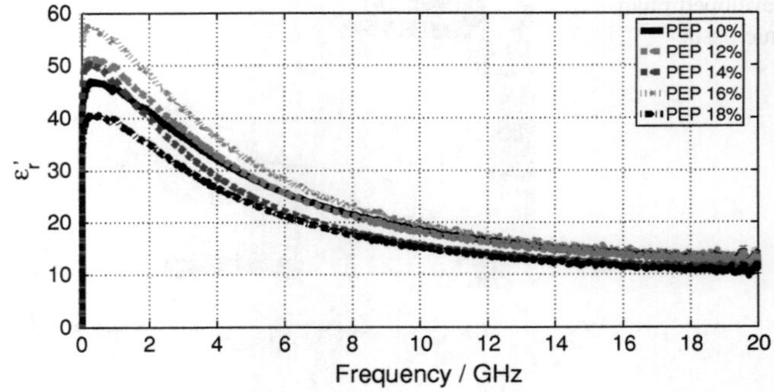

Fig. 6.25 Dielectric constant of phantom for different percentage of polyethylene powder (PEP)

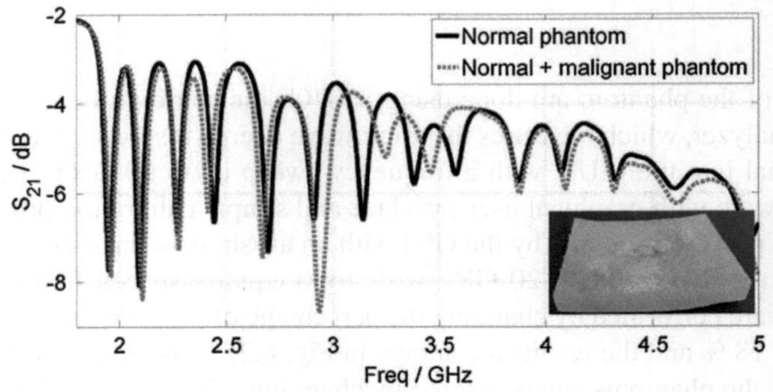

Fig. 6.26 Measurements of a fibro-glandular phantom with and without the malignant tissue. The inset shows a picture of the phantom with parts of malignant tissue (*blue*) [25]

6.3.4 Two-Layered 2D Sensor Array

To simplify the extraction of the dielectric properties, an additional design was made where the feeding microstrip line is located on a separate substrate layer to prevent any influence from it on the interaction between the sensing region of the sensor and the tissue that needs to be analyzed. The principle of operation of this new design is still based on tracking the resonant frequencies of each SRR that will represent 1 pixel of the dielectric image. The new structure design is depicted in Fig. 6.27 and a picture of the prototype is shown in Fig. 6.28.

Measurements with the layer prototype were also made. For the first measurement, a dielectric profile was created where a load of $\varepsilon_r = 2.2$ was placed on the third and seventh ring, a load of $\varepsilon_r = 6$ was located on the fifth ring and a final load of $\varepsilon_r = 10.2$ was placed on the ninth ring. The results are depicted on Fig. 6.29. The extraction methods presented previously were applied and an example of the permittivity model for the third ring is presented on Fig. 6.30. The dielectric values of the perturbers were extracted and are shown in Table 6.6 along with the relative error.

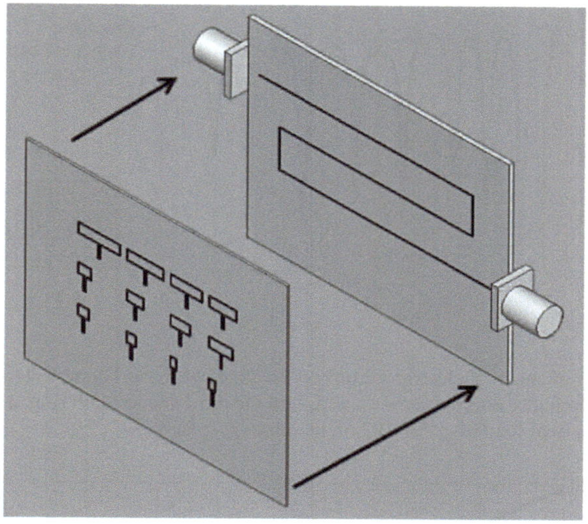

Fig. 6.27 Design of the two-layered 2D sensor array prototype. The microstrip feeding line is located on a separate layer to prevent influence on the sensor [26]

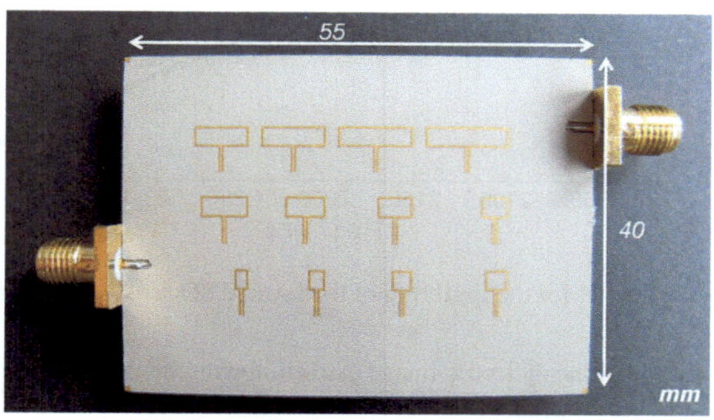

Fig. 6.28 Prototype of the 2D layered sensor array. Dimensions are given in mm [26]

Another profile to test the single cell resolution of the structure was made. A load of $\varepsilon_r = 2.2$ is located on the third and fourth ring and additional loads of $\varepsilon_r = 10.2$ are located on the seventh and eight ring. The results are shown on Fig. 6.31 and the extracted values with the relative error are depicted in Table 6.7. By comparing the results obtained with the larger 1D array presented previously and the extracted dielectric values with both 2D array structures, the simple and the layered, it can be concluded that the extraction methods are less efficient for the 2D array. This result was expected due to the fact that the sensing area of this structure is more difficult to control and the tolerance for the positioning of the reference dielectric bricks is definitely lower. In principle, this is not a main problem since the structure was designed to detect relative dielectric changes and not absolute values. As for a

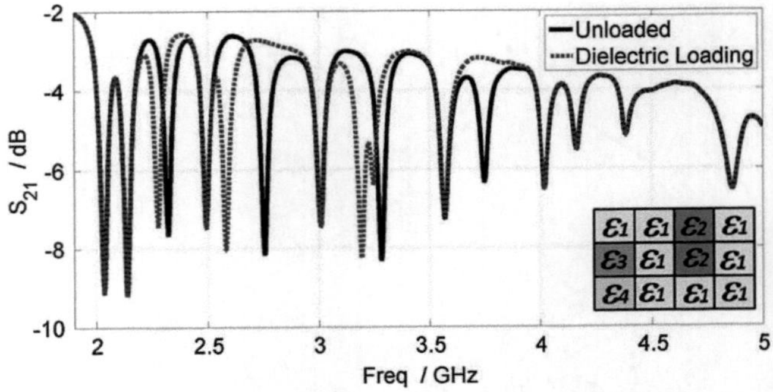

Fig. 6.29 Measurement of dielectric perturbers loading with the layered 2D sensor array. The dielectric profile includes a perturber of $\varepsilon_r = 2.2$ on the third and seventh ring, a load of $\varepsilon_r = 6$ on the fifth ring and a final load of $\varepsilon_r = 10.2$ on the ninth ring [26]

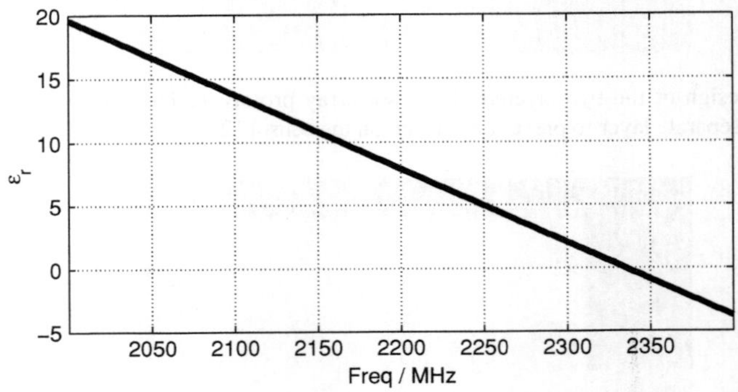

Fig. 6.30 Permittivity model for the third ring of the layered 2D sensor array [26]

Table 6.6 First example of the dielectric image extraction with the layered 2D sensor array

Ring	Original value	Extracted value	Relative error
3	2.2	3.3	0.5
5	6	8.9	0.48
7	2.2	4.7	1.13
9	10.2	15.9	0.55

comparison between the simple 2D sensor and the layered 2D sensor, the last one presents a lower relative error in the extracted dielectric values probably because the feeding microstrip line is not influencing the extraction procedure. In conclusion, both 2D sensor arrays performed well for extracting relative dielectric images but not absolute values. In the next section it will be considered the possibility to use the presented structures not for sensing but for treatment of malignant tissues by means of thermal ablation treatments.

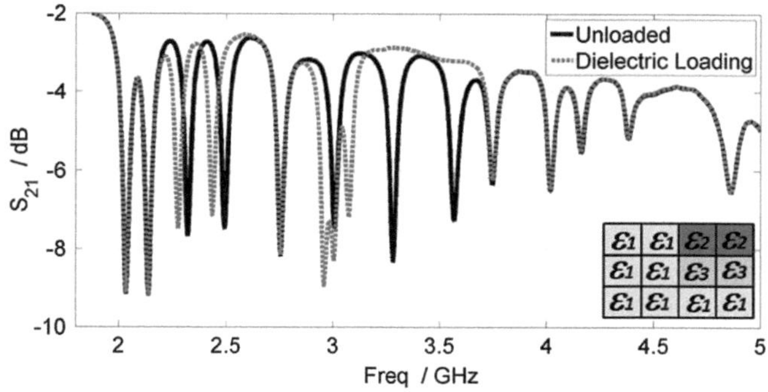

Fig. 6.31 Additional measurement of dielectric perturbers loading with the layered 2D sensor array. A load of $\varepsilon_r = 2.2$ is located on the third and fourth ring and a load of $\varepsilon_r = 10.2$ is located on the seventh and eight ring [26]

Table 6.7 Second example of the dielectric image extraction with the layered 2D sensor array

Ring	Original value	Extracted value	Relative error
3	2.2	3.3	0.5
4	2.2	2.3	0.04
7	10.2	12.3	0.2
8	10.2	17.7	0.73

6.4 Thermal Ablation Therapy

Research and advancements in the electromagnetic field have lead to the development of biomedical devices with various diagnostic and therapeutic applications. There are millions of people around the world who get diagnosed by different types of lung, liver and breast cancers, with primary and secondary hepatic tumors among the most common. In many cases, the patient cannot undergo the classical open surgical methods because of the location of the tumor tissues inside critical body areas. In these cases, thermal ablation is an adequate choice for treatment, since the temperature rise is enough for the size and location of the tumor allowing its eradication. For this reason, thermal ablation therapies have gained interest in the last years, especially for microwave frequencies. The term ablation is defined as destruction or eradication of tissue by extreme hyperthermia (heat). As well as with surgical removal, the rise in temperature is concentrated around the tumor but in a less invasive way, allowing for lower risks for the patient and a faster recovery time [30]. The procedure can be done by using RF, microwaves, laser or ultrasound signals. The most commonly used is RF ablation (RFA) but microwaves are gaining position since they can be used in all the same types of tissues as RFA without the problems with the properties of the tissue, such as conductivity (σ) and changes in impedance. Also microwave ablation can heat faster and with lower input powers. Nevertheless, there are some disadvantages in microwave ablation (MWA) like a lower control of the ablation zone and multiple MWA probes might be necessary to cover the desired ablation zone [31].

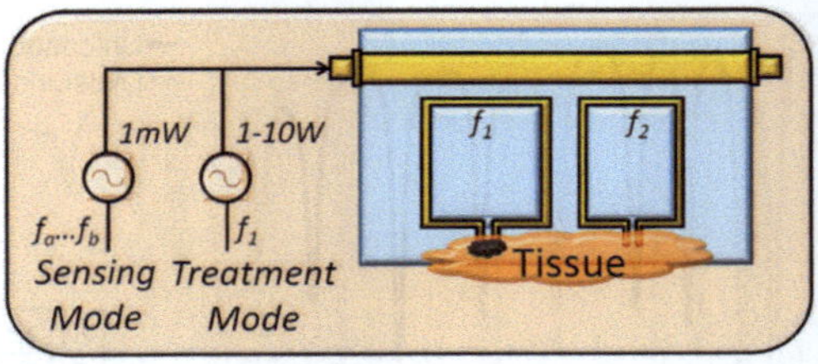

Fig. 6.32 Overview of the dual mode tool for analysis of organic tissue and thermal ablation treatment

In this work, the focus is designing tools operating at microwave frequencies that can perform thermal ablation, where the rise in temperature is well concentrated inside the tumor. The overview of the tool is presented in Fig. 6.32. The novel idea is that the sensor previously presented for analysis of organic tissue can have a second operation mode and perform treatment of malignant tissue in a localized manner by a well focussed rise in temperature of the desired area. This approach of sensing and treatment by thermal ablation with the same device offers several benefits. First of all, the automatic alignment between the determined tumor location and subsequent heat application. Additionally, the tool is ease to use and the time needed is lower than for RFA. Furthermore a control of the ablated area can be done by using the tool in the sensing mode after the ablation procedure is completed.

6.4.1 Thermal Ablation Theory

Thermal therapy includes all therapeutic treatments based on transfer of heat into or out of the body, to achieve effective eradication of tumorous tissue without hurting or damaging the healthy tissues. The extent of ablated tissue is predominantly determined by the applied power and time [32]. Generally, thermal therapy is put into four categories, influenced by the temperature levels and treatment duration. The first is diathermia for physiotherapy applications with temperatures up to 41 °C for rheumatic diseases treatment. The second category is moderate hyperthermia with temperatures between 42 and 45 °C. It is used for cancer treatment in combination with chemotherapy to increase the success rate of the patient. The third category is called thermal ablation and this technique can reach temperatures above 50 °C and directly destroys or eradicates tumor tissues in a localized area [30]. The fourth and final category is cryotherapy where the principle of operation is freezing the tumors to temperatures of as low as −50 °C for a period of 10 min. Cryoablation operates in similar fashion as RFA. In both treatments a probe is used to deliver the desired effect to the tumor tissues through the skin. But rather than killing the tumor with heat, cryoablation uses an extremely cold gas to freeze it. An overview of the categories is seen in Fig. 6.33.

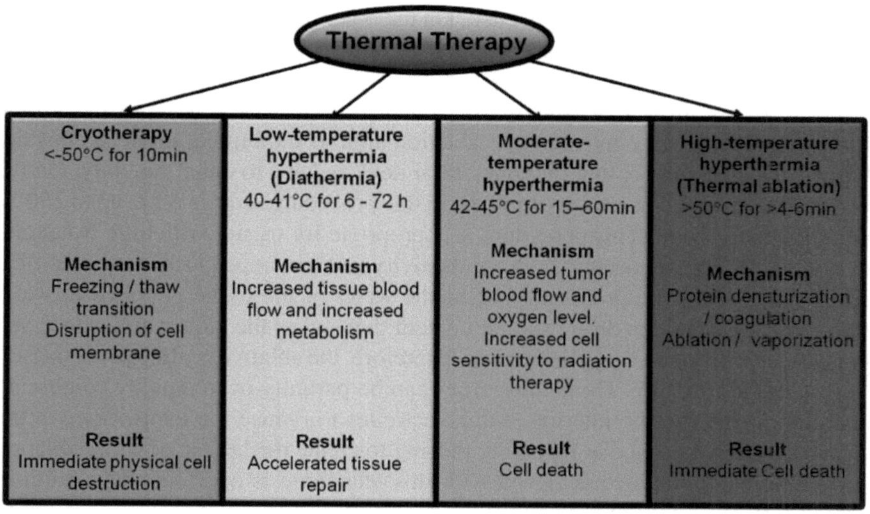

Fig. 6.33 Different thermal therapies and their properties [32]

The main advantage of thermal therapies is to treat the tumor within a localized area where open surgery can be very difficult (e.g., liver) or where organ functionality is needed or desired to be preserved. These kinds of thermal therapies are not suitable for large tumors such as colorectal cancer and some of the cancer types involving skin. The problem areas where tumor boundaries are hard to be defined moderate temperature hyperthermia is more suited than thermal ablation [32]. Thermal ablation is appropriate for treatment in skin, liver, kidney, muscle, and spleen, specially for liver is widely used. Nevertheless, it is also possible to use it in other organs as well, such as the lung and bone [31]. One way to measure the impact of thermal ablation is to control the heat specific absorption rate (SAR) presented in Eq. (6.3). It is shown that a SAR of 1 W/Kg corresponds to an increase of 1 °C in human body temperature [33].

As the thermal ablation therapy with microwaves is the main focus of this study, a comparison with its biggest competitor RFA will be discussed in terms of their limitations and advantages for various tissues, organs and applications. RFA is based on applying electrical current to the body. Tissues undergo conduction because of ionic fluids. As they are not perfect conductors, resistive heating phenomena occurs. Most of the direct heating is produced in a few millimeters area around the RF applicator (electrode), which increases the large ablation zone due to thermal conduction over time. RF currents can be applied in monopolar and bipolar modes. In monopolar mode, a single interstitial electrode is inserted in the body and is used to deliver heat, while ground pads on the skin surface complete the electrical circuit. In bipolar mode current is delivered via two interstitial electrodes. Bipolar mode results in more effectively localized heating between the electrodes. In bipolar mode no background conductivity or ground pads are needed. On the other hand, the monopolar mode has the advantage of ablating a wider area since it requires the use of electrodes connected to the skin surface [31]. Although the results are promising, RF ablation faces many disadvantages. For example, the use of ground pad electrodes on the surface of the skin results is a lower control on the ablation area. If the bipolar mode is consider than the disadvantage is that two interstitial devices have to enter the body. On the other hand, the power necessary to perform this procedure is quite high, up to 250 W with 2.3 A of current. This procedure is appropriate for tissues with high water and ion content that translates into high conductivity and low impedance to the electrical current path. Once the procedure begins the water of the tissue is boiled to water vapor creating a sudden drop in water content that spikes the circuit impedance and translates in a drop of the applied power. Therefore, the ablative heating process itself is a limitation for RFA. These difficulties can be partially overcome by combining a cooling process in the interior of the electrodes to reduce the temperature at the electrode-tissue interface or by power pulsing to create the largest possible ablation zone. The maximum size achieved with this technique is approximately 3 cm in diameter. When a multiple electrode device is used, this can be raised to 5 cm in diameter. As summary, even with high powers, RFA is limited by the reliance on electrical current conduction. High impedance tissues, including the ablation zone itself, preclude effective RFA when the power applied is limited, since high power generators have not been clinically deployed. Therefore, investigations on alternative methods for ablation are necessary [31].

The natural evolution of RFA is of course the use of microwaves. For this approach an interstitial antenna is used, which means that the need of a ground electrode does not apply anymore and the difficulties with electrical current conduction are minimized. Microwaves will propagate through all types of tissue, including water vapor and dehydrated or desiccated tissue inside ablation zones. In the design of the MWA tool is necessary to consider the effective radiation into the surrounding tissue to maximize energy delivered and to control the radiation pattern in order to produce the desired shape of the thermal ablation zone. The mechanism used by MWA is dielectric heating. This phenomenon occurs when an alternating electromagnetic (EM) field is applied to a dielectric material. The heating in the tissue is produced due to the oscillation of water molecules in an out of phase and the EM energy is then absorbed and converted into heat. With tissues that contain a high water content the absorbtion is higher (e.g., most solid organs) than in tissue with low water content (e.g., fat). For MWA the key properties are conductivity, complex relative permittivity and thermal conductivity of the organic tissue that are influenced by the electrical, thermal and mechanical properties of the tissue. One last parameter that has to be taken into account is the rate of blood perfusion that will determine the size of the ablation zone because it will behave as a heat sink. Another reason why thermal ablation is a good choice of treatment is that dielectric properties of cancer cells are 10–30 % higher than healthy tissues, and the water content is also different allowing the cancer cells to get heated more easily than normal tissues. The reasons for this difference are that normal tissue has about 2–10 times more blood flow than tumor tissue. As a result, the temperature change in healthy tissue differs remarkably from tumor tissue.

MWA has of course a few disadvantages. It may required input powers of up to 30 W, and therefore, the coaxial cables used to feed the antenna can malfunction due to heating, since they are very small in diameter, mainly because they have to enter the body through minimal invasive incisions. Thicker cables are not adequate, since they cannot be used for percutaneous application. By limiting the power and time these difficulties can be overcome, but the size of the ablation zone is reduced. Another possibility, as with RFA, is to cool the antenna by using water or cryogenic gas. There has been considerable research and clinical studies, for example in [34] a design of a microwave applicator for MWA is described, which can ablate localized zones up to $3 \times 2.5 \times 2.5$ cm^3 with 60 W of power for 2–4 min. It is also used in conjunction with radiotherapy or chemotherapy, as it is evident that hyperthermia can increase the efficiency of these therapies [31].

It is important to analyze if the previously used SRRs are able to give an adequate temperature rise to actually kill the cancer cells. With temperatures up to 41 °C no real long-term effect can be achieved. Above 46 °C the irreversible damage in the cell begins. From 46 to 52 °C, less time to kill the cells is needed, and above 60 °C, there is no possibility for the cell to recover and water vaporization, desiccation and carbonization occurs [31]. The tool that will be presented here for certain scenarios is able to rise the temperature well beyond 60 °C, and therefore, is quite suitable for thermal ablation.

6.4.1.1 Bioheat Transfer in Organic Tissues

Heat transfer in organic tissues is very different to heat transfer in materials that are not alive. It is a set of complex multiple physical phenomena that can be well explained by the bioheat transfer process as seen in Fig. 6.34. One of the most important mechanism to be consider is the forced convective heat transfer produced by the blood perfusion. Blood perfusion is the continuous blood flow through tissues and organs via a complex network of branching vessels. This phenomena has to be carefully consider in a thermal ablation therapy to prevent healthy tissue distant from the malignant tissue to be heated and destroyed. Therefore, several parameters influence the transfer of heat in tissues such as vascular geometry or blood temperature (arterial and venous blood temperatures are different), among others. Other physical phenomena that also influenced heat transfer in biological tissues besides convection are conduction, radiation, metabolism heat generation, evaporation and phase change [35]. Pennes proposed a model to describe the effects of metabolism and blood perfusion on the energy balance within tissue. These effects were incorporated into the standard thermal diffusion equation, which is written in its simplified form as [36]

$$\rho c \frac{\partial T}{\partial t} = \nabla k \nabla T + SAR - \rho_{bl} c_{bl} w_{bl} (T - T_{bl}) \tag{6.6}$$

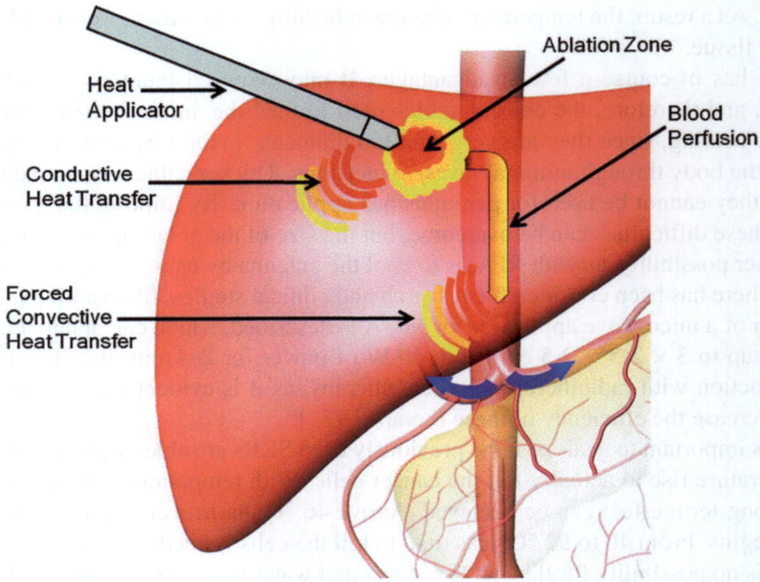

Fig. 6.34 Bioheat transfer mechanisms in organic tissue. Conduction and forced convection due to blood perfusion are highlighted

where ρ is the tissue density (kg/m^3), c is the specific heat capacity (J/kg · K), ρ_{bl} is the blood density (kg/m^3), SAR is the specific absorption rate, c_{bl} is the specific heat capacity of blood, w_{bl} is the blood perfusion, T_{bl} is the blood temperature (K) and k is thermal conductivity (W/m · K). This equation is typically known as the "Bioheat equation" and several assumptions must be taken into account to validate the model. The metabolic heat generation is assumed to be homogeneously distributed throughout the tissue of interest as rate of energy deposition per unit volume. The blood perfusion effect is homogeneous and isotropic. Thermal equilibration occurs in the microcirculatory capillary bed, therefore is assumed to be no energy transfer before or after the blood passes through the capillaries, so that the temperature at which it enters the venous circulation is that of the local tissue. The total energy exchange between blood and tissue is directly proportional to the density ρ, specific heat capacity (J) and blood perfusion rate w_{bl}. This thermal transport model is analogous to the process of mass transport between blood and tissue, which is confined primarily to the capillary bed [35]. Many researchers have used this model in the last decades and opinions are contradictory in terms of the validity of the model. Nevertheless it is the most widely used model to described heat transfer in biological tissues and in many particular scenarios it was proven valid. For analyzing thermal ablation with the sensors presented in this work, specially in simulations using the CST Microwave Studio software, the bioheat equation will be used since it considers all the important phenomenons for the application at hand.

6.4.2 Dual Mode Tool Design Process

The main goal is to extend the design of the structures presented in the previous section to operate in two modes, sensing and treatment. The first mode is in charge of extracting the dielectric properties of the tissues and the second mode performs thermal ablation treatment when a malignancy is located. A reduction in size is necessary for this final prototype so it can be included in a minimal invasive surgery tool. Once an abnormality is detected, the tool can change its operation mode and heat the tissue at a particular location, i.e. around one specific SRR of the structure and the hot spot can be localized around the gap of this certain SRR. The tissue under test (TUT) is heated to the required temperatures to perform thermal ablation treatments. The most commonly used tool in MWA is a mechanical bladder type needle. These needles have space at the tip to put a microwave applicator of certain maximum dimensions. The sizes of the needles can be varied for location and method of treatment; usually the diameter and length varies between 5 mm and 1 cm. For the design of the dual mode tool, the operation frequency and size of the tool is critical, as the size of the SRRs depends upon the frequency of operation, the idea is to investigate the temperature distribution at different frequencies and sizes of the SRRs without exceeding the maximum dimensions of the needle microwave applicator. Since the tissues properties depend on the operating frequency, the study is also conducted to see the temperature distribution and size of the ablation zone for different operating frequencies i.e. sizes of the sensor. All the major design parameters are mentioned in Table 6.8.

After several optimization steps in the design process, four prototypes were selected as first proof of concept. The first two prototypes have one SRR with different geometries, circular and square. Additionally, a prototype with three circular SRRs and another with four square SRRs were also developed. The frequency and size parameters were accomplished, since the structures have an operating frequency range of 4–8 GHz and a size no larger than the requirement as will be seen in the pictures of the prototypes. The typical teflon isolation layer was again included and for the structures with more than one ring, they are decoupled to resonate at different frequencies by changing the width or diameter of the rings.

In Fig. 6.35 are shown two views and parameters of the design. Figure 6.35a shows the cross-section where the substrate, ring and microstrip line for excitation are discernable. In Fig. 6.35b a full view of the teflon protection layer with its dimensions is depicted.

For the treatment mode of operation, extensive thermal simulations were performed. Simulations were done using CST Microwave Studio to monitor the E field and specific absorption rate (SAR) at the frequency of operation. The thermal simulations can represent stationary or transient temperature distributions. In our case the thermal source used was a thermal loss distribution computed from electromagnetic fields and then included into the thermal simulation. Depending of the requirements of the model many different thermal aspects may be considered. The most interesting feature about this solver is that biological heating/cooling mechanisms such as blood

Table 6.8 Design parameters of the thermal ablation tool

Center operation frequency	5 GHz
Dimensions (sensor)	Length = 1 cm; Width = 5 mm; Height = 3 mm
Number of SRRs	4
Isolation layer	Teflon
Temperature measurement	Infrared camera; Platinum RTD

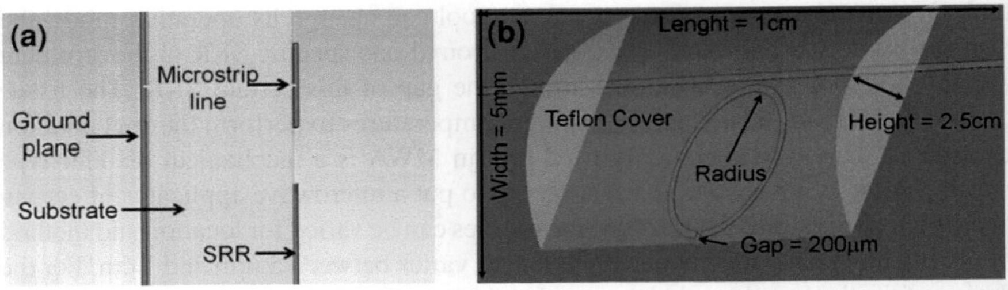

Fig. 6.35 Sensor design: **a** Cross-section; **b** Tilted view with circular SRRs [29]

perfusion and basal metabolic rate can be taken into account. It is also possible to do coupled simulations in combination with other solvers. The use of Voxel models is also supported.

The first step is to calculate the loss monitor for the specific resonance frequency of the SRR that will be used in the heating process by using the electromagnetic solver. After the losses have been monitored, one can select between the thermal stationary or transient solvers and visualize the temperature distribution using the Bioheat equation. As the final measurements are done on physical phantoms at room temperature, the background temperature is set to 25 °C. The background material is set as air.

The fundamental purpose of the simulations is to verify the rise in temperature as power increases and approximate the thermal lesion size which can be ablated at a given frequency and input power. Investigations have also been made for different sizes of the SRRs and their effects on temperature and lesion size. To simplify the simulations, the tissue was consider homogeneous and no malignant section was considered. For future work, the effect of different types of malignant tissues with different sizes and locations within the healthy tissue can be addressed specially to proof that malignant tissue can heat faster than its healthy counterpart. For the simulations, lung and liver tissue were used and Table 6.9 has their dielectric properties.

In Fig. 6.36 is an example where the structure with one square ring has an input power of 5 W and a cross section of the temperature distribution is shown. The maximum obtained temperature around the hot spot located in the tissue surrounding the gap of the ring is more than 150 °C and an approximate size of the ablation zone is depicted. For an adequate design several parameters needed to be analyzed in relation with the size of the ablation zone such as power, size of SRRs and size of the tissue itself. The results will be discussed.

Table 6.9 Dielectric properties of liver and lung tissue at 10 GHz

	Liver	Lung
Epsilon	69.82	31.64
Electrical conductivity (s/m)	0.4872	0.3057
Thermal conductivity ($W/K \cdot m^{-1}$)	0.469	0.624
Heat capacity ($kJ/K \cdot kg^{-1}$)	3.6	3.6
Diffusivity (m^2/s)	$1.222e^{-7}$	$1.6999e^{-7}$
Bloodflow ($W\ K^{-1}\ m^{-3}$)	68,000	9,500
Metabolic rate (W/m^3)	12,000	1,700

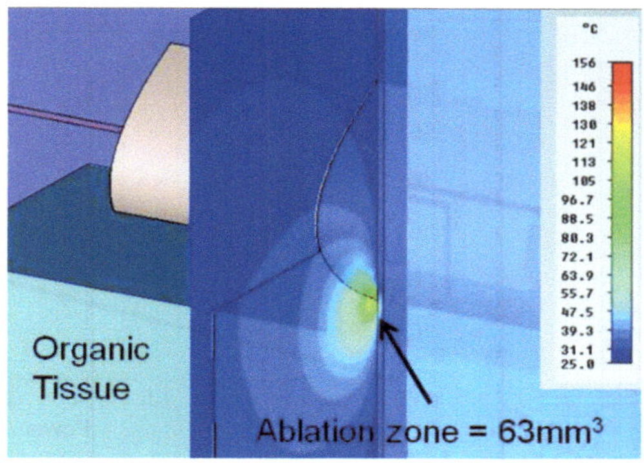

Fig. 6.36 Cross section of temperature distribution inside tissue. The input power is 5 W and the lesion approximate size is depicted [29]

6.4.2.1 Effect of Power on Temperature and Lesion Size

The first analysis to be made was the effect of the input power on the maximum temperature that could be achieved and size of the ablation zone. In Table 6.10 is shown how temperature rises when the input power increases for the 1-square-SRR. The SRR from this structure has its resonance frequency at 5.58 GHz. The lesion size represents the volume boundary of tissue where the temperature reaches its maximum value and keeps a temperature within 5 % of this maximum value. Simulations show that a temperature of above 110 °C can be achieved with only 5 W of input power and a lesion size of $4.8 \times 3.3 \times 4\ mm^3$. When compared with RFA and other MWA techniques, this is a very power efficient tool and can be used for tumor sizes of up to 1 cm. For larger tumors a tool with more SRRs can be used, the limit is the amount of SRRs that can be adequately controlled in terms of their resonance frequencies not overlapping each other when disturbed by the TUT. In Fig. 6.37 is presented another approach to see the results where the volume of the lesion is given in terms of the input power. The volume of the ablation zone increases and can be fitted with a quadratic polynomial. The increase in temperature and ablation zone can also be seen graphically in Fig. 6.38 where the cross-section of the sensor is shown along its correspondent temperature distribution at different input powers with a maximum reached temperature inside the tissue near the gap of the SRRs.

Table 6.10 Simulation results for the 1-Square-SRR prototype

Power (W)	Max. temperature (°C)	Lesion size (mm)	Volume lesion (mm^3)
0	26	0 × 0 × 0	0
1.5	54	2.5 × 2 × 2.3	11.5
2.5	75	3 × 2.4 × 2.8	20.16
3.5	93	3.8 × 2.9 × 3.2	35.26
4	102	4 × 3 × 3.6	43.2
4.5	117	4.8 × 3.3 × 4	63.36

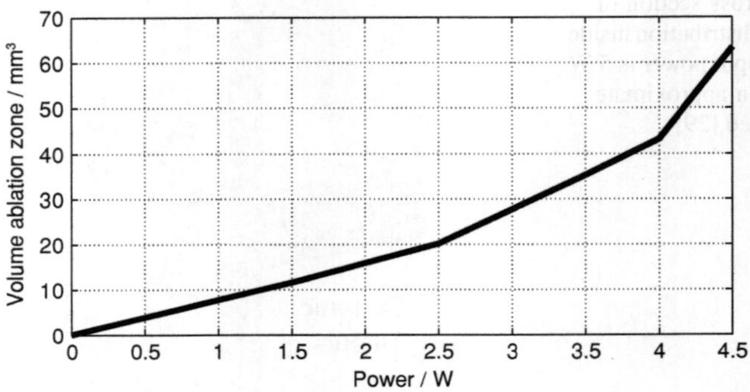

Fig. 6.37 Volume of the lesion versus input power for the 1-Square-SRR prototype

The same test was made with all of the constructed prototypes and although they have similar behavior (temperature increase versus power increase) according to its frequency of operation, the reached temperature and size of lesion for a specific input power is slightly different for all SRRs.

6.4.2.2 Effect of Ring Size on Temperature and Lesion Size

This section investigates the effects of the SRR size on the temperature distribution and thermal ablation lesion size. In previous simulations the size of the ring was fixed for temperature analysis. In this section the width or radius of the SRRs are changed depending on the structure being analyzed and the effects are summarized for one particular structure in Table 6.11. The width of the square single ring is changed from 4 to 8 mm with a constant input power of 5 W and the structure is loaded with lung tissue. All the other parameters like gap of ring, width and thickness are kept the same.

From the simulation results can be seen that by keeping the power to the same level and increasing the size of the SRR the lesion size increases. From a theoretical point of view this is clear, as the size increases the frequency of operation decreases. At low frequency the EM energy can penetrate more into the tissue and make a larger

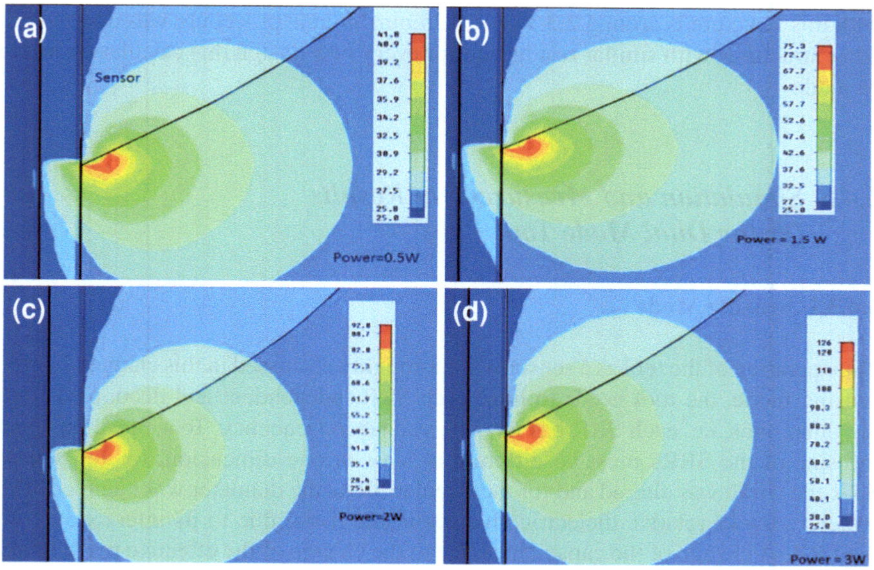

Fig. 6.38 Temperature distribution versus power (side view) for the 1-Square-SRR prototype. **a** Power = 0.5 W, Max. Temp. 92.8 °C; **b** Power = 1.5 W, Max. Temp. 75.3 °C; **c** Power = 2 W, Max. Temp. 92.8 °C; **d** Power = 3 W, Max. Temp. 126 °C

Table 6.11 Effect of the size of the SRR for the 1-square-SRR prototype

Width-SRR (mm)	Frequency (GHz)	Max. Temperature (°C)	Lesion size (mm)	Volume lesion (mm^3)
4	3.7	134	5.5 × 2.3 × 4	51
6	2.8	135	7 × 4 × 4	112
8	2.5	141	9 × 5 × 5	225

ablation zone at the same power level. Nevertheless it has to be taken into account that the tissue properties are different at different frequency and the size of the tool has to satisfy the maximum size allowed to be included in a minimal invasive tool.

6.4.2.3 Effect of Tissue Dimensions and Distance of Tissue from Sensor

A final analysis was performed to see the influence of the size and position of the tissue in the behavior of the tool. It is important to maximize the size of the ablation zone while minimizing the time and input power for maximum efficiency. When the SRR is loaded with tissue an increase in the total ring capacitance occurs resulting in a shift in the resonance frequency. For effective use of the tool, it is important to have an appropriate coupling distance between the tool and the target tissue. The simulation results indicate that the tissue cannot be farther away than 350 μm for the structure with two square SRRs and the maximum size of tissue that can be covered with this structure is around 2.3 × 1 × 0.25 mm^3. Same procedure was done for all other structures with similar results and for every new design this calculation has to be made.

6.4.3 *Simulation and Measurement Results for the Dual Mode Tool*

6.4.3.1 Sensing Mode

The operation of the tool as a sensor was thoroughly discussed in this chapter. For the sensing mode, the tool needs to have each SRR independent and decoupled from the other ones, i.e. each SRR has its own resonance frequency. To achieve this goal the size of the SRRs must be different at least in one dimension, for the square SRR the width was altered and for the circular SRRs the diameter was changed. The structure tracks relative dielectric changes in the tissue due to its interaction with the SRR. A change of the capacitance due to the change of the effective permittivity is produced and therefore differences on the tissues, such as abnormalities, can be detected. With this information a dielectric image of the TUT can be obtained with a resolution, i.e. amount of pixels, equal to the number of SRRs. The used power in this mode of operation is 1 mW and represents no danger to the tissue of being heated.

6.4.3.2 Treatment Mode

The second mode of operation i.e. the treatment mode will be addressed more extensive. After extensive thermal simulations were done to understand all important parameters of the structure the final designs were made and further thermal simulations were done on these designs to make sure the performance was adequate before fabricating the structures. In Fig. 6.39a the results of the thermal distribution for the structure with three circular SRRs are shown. Figure 6.39b shows the scattering parameters and the resonant frequencies are clearly independent and unique. It is clear that the hot spot is located around the second ring by selecting the appropriate resonance frequency and the temperature at this point rose to 50.6 °C with an input power of 5 W. With the same power, the first SRR of the structure manages to rise the temperature to 51.4 °C and the third SRR to 50.5 °C. In Fig. 6.40 the relation between the power and the temperature obtained at the different hot spots for this structure is shown. For the structure with square rings the same procedure for the thermal analysis was made. The results of the thermal distribution for the structure with four square SRRs are similar as senn in [37]. The second ring was used for the ablation treatment and reached a temperature of 74 °C with an input power of 10 W. In Fig. 6.41 the relation between the power and the correspondent

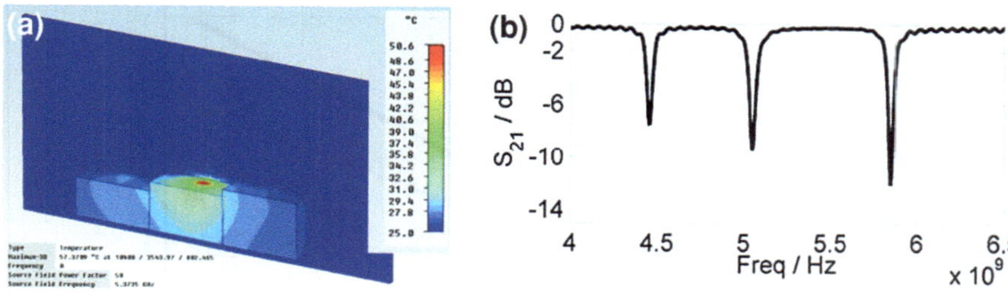

Fig. 6.39 **a** Simulated model and thermal results for the structure 3-circular-SRR. **b** S_{21} parameters [37]

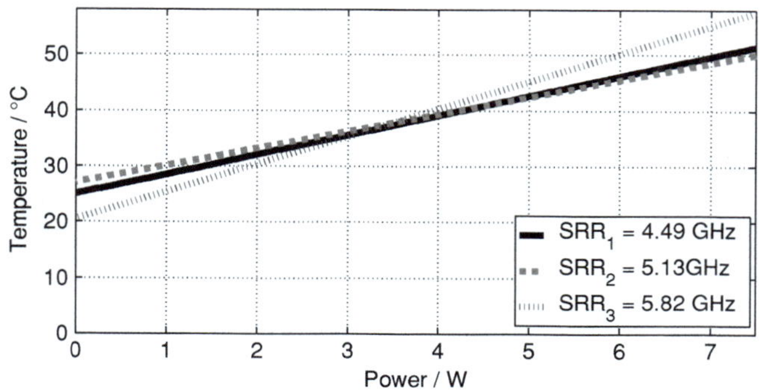

Fig. 6.40 Comparison between the maximum temperature reached with a certain input power with the 3-circular-SRR structure

temperature obtained at the hot spots for this structure is shown. Although all SRRs for both structures presented do not heat at exactly the same temperature while being excited by the same input power the increase in temperature is quite similar. This can be seen by calculating the slope of each SRR curve in Fig. 6.40 and the value vary between 3 and 4 meaning that in average an increase of 3.5 °C/W can be achieved with any SRR of the presented structures. From the simulation results it was derived that the structures were very promising for thermal ablation treatment and therefore the fabrication of the prototypes to proof the concept was done and the results will be now discussed.

6.4.3.3 Prototypes

In Fig. 6.42 the four designed structures are shown. They were constructed on a Rogers 6010 substrate with $\epsilon_r = 10.2$ and thickness of 254 μm. The dimensions of the structures are depicted and it should be emphasized that a reduction in size for an actual tool that can be used on real patients is possible with the tradeoff of a higher frequency of operation and higher sensitivity to the imminent losses from the biological tissues.

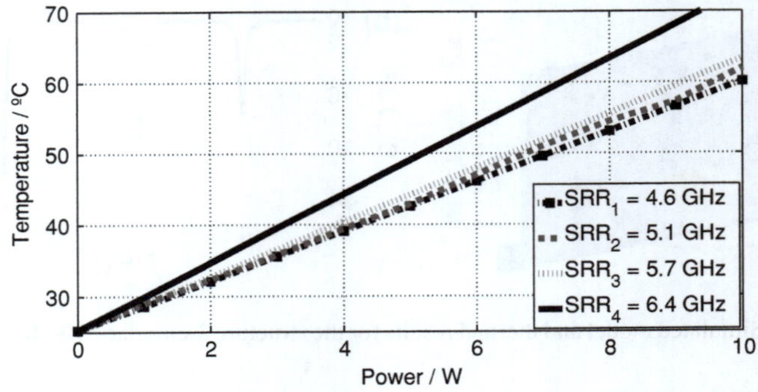

Fig. 6.41 Comparison between the maximum temperature reached with a certain input power with the 4-square-SRR structure [37]

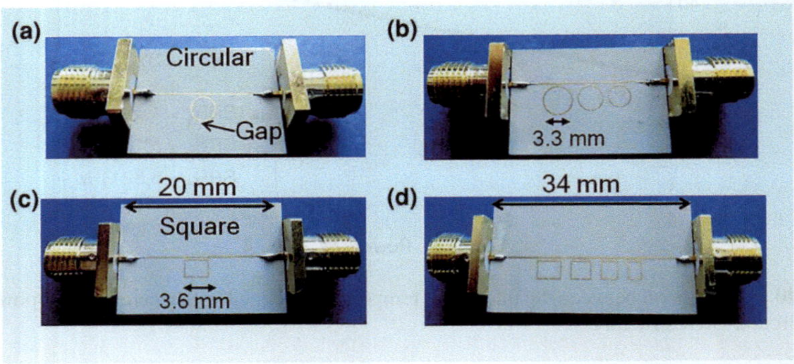

Fig. 6.42 Developed prototypes: **a** One circular SRR; **b** Three circular SRRs; **c** One square SRR and **d** Four square SRRs [37]

The parameters of the fabricated prototypes like frequency of operation, dimensions of rings and Teflon as protection layer are mentioned in Tables 6.12 and 6.13 and are in good agreement with the goals set for this structures given in Table 6.8. In Sect. 2.3.2, it was discussed that for a microstrip based structure, the square rings are more appropriate since the coupling surface between the feeding microstrip line and the SRR is larger. For coplanar based devices, the circular rings are more appropriate in terms of coupling. Nevertheless, both cases were designed and fabricated for the microstrip design to confirm which geometry performs better for the specific target application of thermal ablation.

6.4.3.4 Measurement and Results

Measurements are done using the phantoms of human tissue discussed previously. The first step is to operate the tool at the sensing mode to identify where the

Table 6.12 Parameters of the square SRR based prototype

Parameters	1-Square-SRR	4-Square-SRR
Frequency of operation	$f = 5.2$ GHz	$f_1 = 4.8$ GHz; $f_2 = 5.4$ GHz; $f_3 = 6.1$ GHz; $f_4 = 6.7$ GHz
Dimensions rings	$L = 3.6$ mm; $gap = 0.2$ mm	$L_1 = 4.1$ mm, $L_2 = 3.5$ mm, $L_3 = 2.9$ mm, $L_4 = 2.4$ mm; $gap = 0.2$ mm
Isolation layer	Teflon	Teflon
Temperature measurement	Infrared camera; Platinum RTD	Infrared camera; Platinum RTD

Table 6.13 Parameters of the circular SRR based prototype

Parameters	1-Circular-SRR	3-Circular-SRR
Frequency of operation	$f = 5.9$ GHz	$f_1 = 4.7$ GHz, $f_2 = 5.5$ GHz, $f_3 = 6.3$ GHz
Dimensions rings	$r = 1.6$ mm; $gap = 0.2$ mm	$r_1 = 1.9$ mm, $r_2 = 1.7$ mm, $r_3 = 1.5$ mm; $gap = 0.2$ mm
Isolation layer	Teflon	Teflon
Temperature measurement	Infrared camera; Platinum RTD	Infrared camera; Platinum RTD

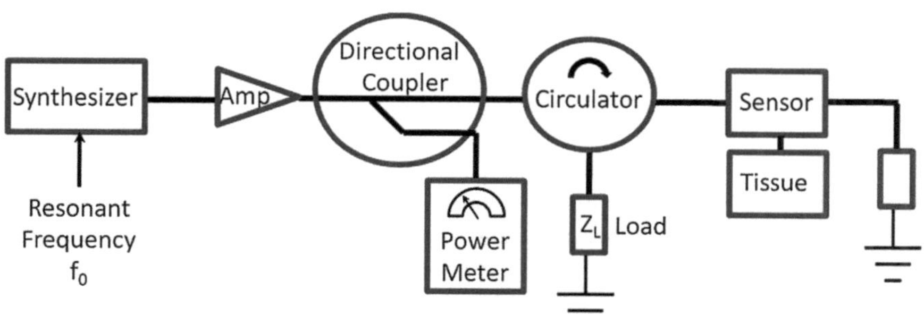

Fig. 6.43 Measurement setup for thermal ablation of phantom tissue [37]

abnormality is located. After the shifted resonance frequency of the loaded structure is obtained, the tool is switched to the treatment mode and the input power is slowly increased to 10 W. For the measurements a special setup is constructed specially to be able to operate the sensor in the treatment mode. The setup is depicted in Fig. 6.43. The resonant frequency of the SRR that will perform the heating is given as an input parameter to the synthesizer. The signal is then amplified to the appropriate power which is controlled manually and displayed in a power meter thanks to the connection of a directional coupler. Then a circulator is used to protect all the equipment from undesired reflected signals and finally the amplified signal is delivered to the structure and the heating of the tissue around the desired SRR is performed.

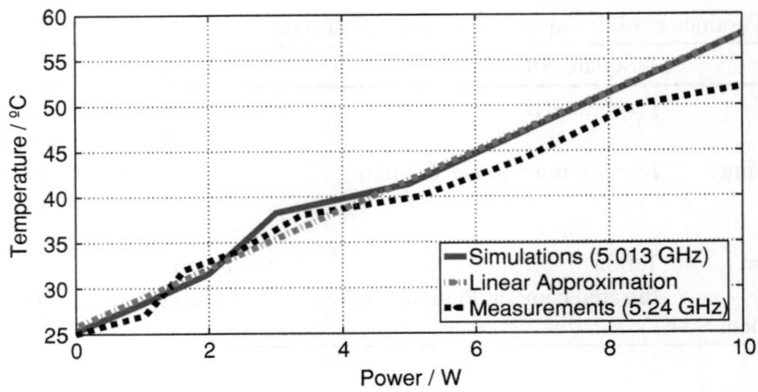

Fig. 6.44 Temperature measurements versus simulations for the 1-square-SRR [37]

Table 6.14 Temperature measurements of 1-square-SRR prototype with phantom tissue. Operating frequency is 5.24 GHz

Power (W)	Max. temperature (°C)
0	25
0.51	26
1.03	27
1.63	32
2.42	34
3.4	38
5.16	40
6.68	44
8.4	50
10	52

The most important parameter to measure and control is the temperature. There are four types of simple temperature sensors available in the market, thermocouples, resistance temperature devices (RTD), infrared camera based temperature measurement devices and optical fiber based. Thermocouples consist of two different types of metal wires or stripes joined together at one end. The temperature changes at the junction and induces an electromotive force (EMF) to the other end. Resistance temperature devices (RTD) work on the principle that electrical resistance of the material changes with respect to its temperature and resistance rises linearly with temperature. Infrared temperature sensor are non conducting devices, and temperature is measured by thermal radiation emitted by the material. Although in clinical procedures of thermal ablation optical fiber based temperature sensors are preferred since they are immune to microwave radiations, for this initial prototype a platinum thin film resistive temperature detector (RTD) and an infrared camera were used. The disadvantage is that they can only detect surface temperature which is less than the inside temperature of the tissue, for this reason a difference between simulations and measurements can be seen but a good trend in the rise of temperature is obtained.

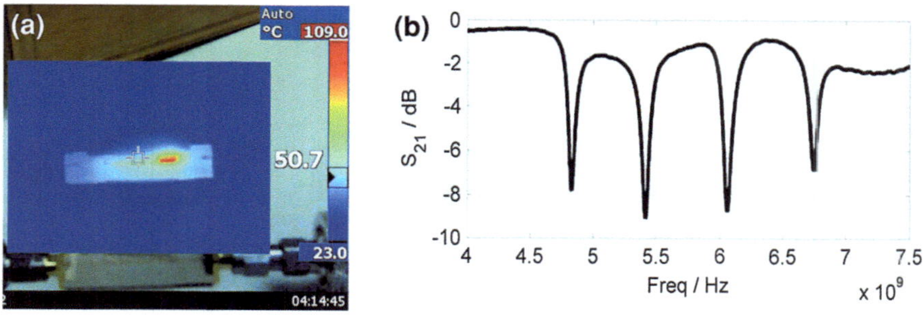

Fig. 6.45 **a** Thermal measurements of the structure with 4-square-SRRs. **b** S_{21} parameter [37]

Table 6.15 Temperature measurements of 4-square-SRR prototype with phantom tissue. The first SRR was used at a frequency of 4.78 GHz

Power (W)	Max. temperature (°C)
0	27
0.6	41
1.46	53
2.2	61
3.66	75
4.67	83
6.43	96
8.20	103
10	109

The first step is to compare between the measurements and simulations. This comparison can be seen for the 1-circular-SRR prototype in Fig. 6.44. The difference is mainly due to the fact that the absolute dielectric value obtained from the used phantom can contain small errors and the influence from the measurement setup components i.e. the losses inserted by these elements is unknown. The measured temperature distribution for this structure is summarized in Table 6.14. The temperature reached 52 °C with 10 W of input power. The obtained values are slightly lower than what was expected from the simulations. The reason is than in simulations the metallization of the structures was done using perfect electric conductor material (PEC) and the actual structures are made of gold. The material PEC does not include information about the thermal conductivity of the material. Although the CST microwave software recommends to to the thermal simulations with PEC for future designs it is better to select the correct material that will be used for the metallization of the structures.

Another example is depicted in Fig. 6.45. Here the 4-square-SRRs structure was used and the malignant tissue around the first ring was subjected to thermal ablation. The temperature reaches 109 °C with 10 W of input power. The Fig. 6.45b shows the scattering parameter S_{21} where four independent resonant frequencies are depicted. In Table 6.15 is a summary of the obtained temperatures at different power levels. An appropriate rise in temperature to perform thermal ablation treatment can be seen.

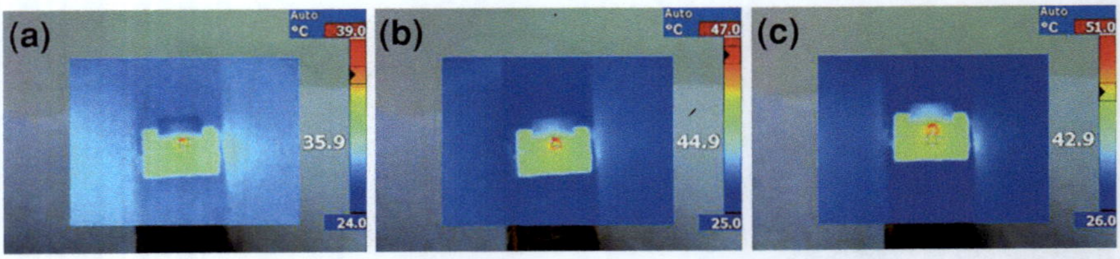

Fig. 6.46 Measurement results for the 1-square-SSR prototype at different input power levels. **a** Power = 2 W, Max. Temp. 39 °C; **b** Power = 3 W, Max. Temp. 48 °C; **c** Power = 4 W, Max. Temp. 51 °C

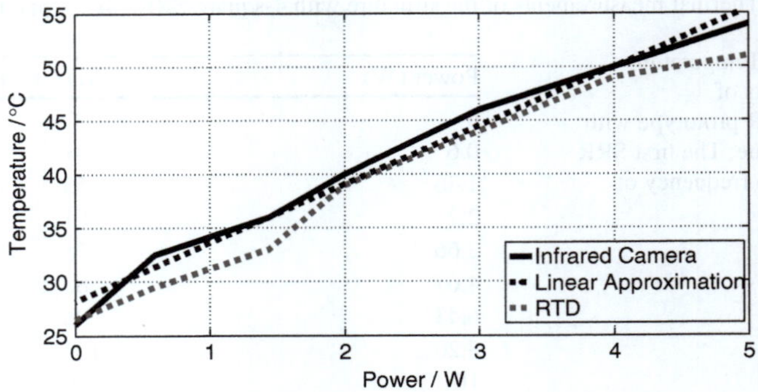

Fig. 6.47 Measurement results power versus temperature for the two square SRR prototype. The second ring was used for the thermal treatment [29]

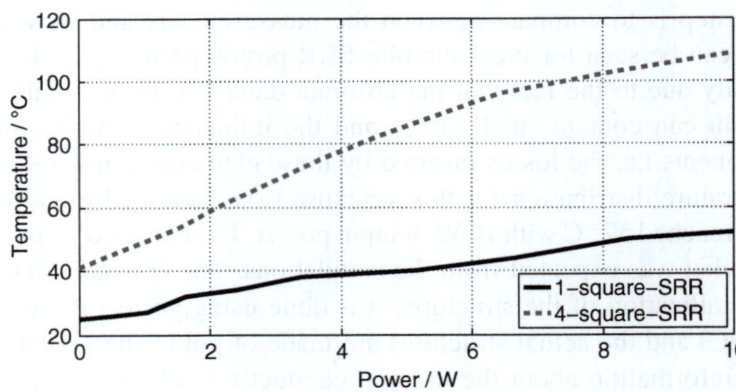

Fig. 6.48 Summary of the obtained temperatures at different power levels for the structure with 1-square-SRR and 4-square-SRR [37]

A final example is depicted in Fig. 6.46 where several pictures were taken with the infrared camera of the structure with 1-square-SRR when subject to different input powers. The progressive rise in temperature can be clearly seen and with 4.5 W a temperature of 51 °C was reached. In this figure can also be appreciated the effect of the conduction phenomena where the heat propagates in the phantom and does not

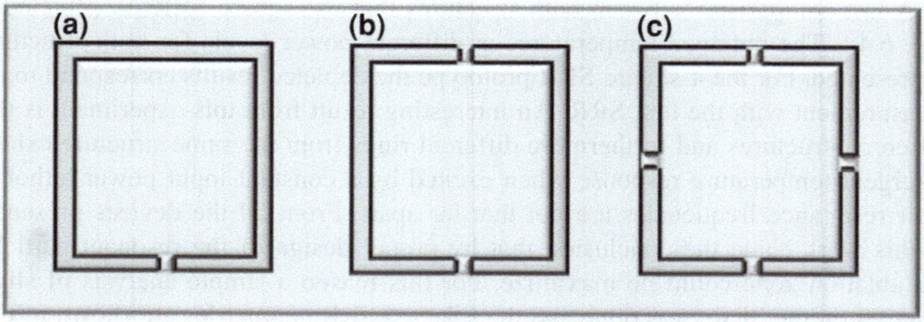

Fig. 6.49 Different analyzed SRR particles. **a** 1-gap-SRR design. **b** 2-gap-SRR design. **c** 4-gap-SRR design

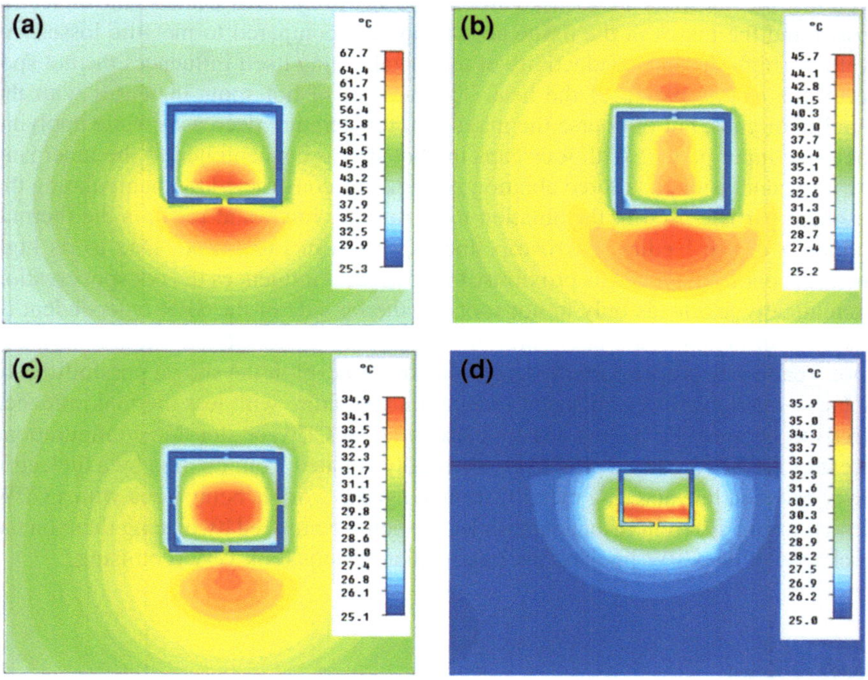

Fig. 6.50 Thermal simulation results. **a** 1-gap SRR, **b** 2-gaps SRR, **c** 4-gaps SRR, **d** Microstrip excited 1-gap SRR

remain confined to the gap of the heating SRR. This has to be carefully considered in a final design for human use, especially because combined with the forced convective heat transfer produced by the blood perfusion healthy tissues far from the location of the tool can be ablated. In Fig. 6.47 it is shown a comparison between the results obtained with the RTD and the infrared camera for the 2-square-SRRs device to make sure that both methods measure similar values and a good agreement was achieved.

A last comparison between both structures that use square SRRs is depicted in Fig. 6.48. The obtained temperatures at different power levels for both structures is presented. For the 4-square-SRR prototype the depicted results correspond to the measurement with the first SRR. An interesting result from this experiment is that different structures and furthermore different rings from the same structure exhibit divergent temperature response when excited by a constant input power although their resonance frequencies are not that far apart. From all the devices presented in this work came the conclusion that by proper design of the resonant particles the ablation zone could be maximize. For this reason a simple analysis of slight changes on the SRRs was done and the selected SRR geometries are shown in Fig. 6.49. To analyze only the behavior of the SRR without any disturbances, the feeding microstrip line of the original structure is replaced by a discrete port on the gap. The temperature distribution for four different scenarios is shown in Fig. 6.50. It is seen that the hot spots are focused around the gap of the SRR because the dielectric losses and the losses in the tissue are dominating compared to metallic losses. As well, the high thermal conductivity of the metallic ring itself influences the hot spot form due to dissipation of the heat. The number of hot spots increases with the number of gaps, but of course the maximum temperature decreases while applying the same input power. With four gaps the hot spot is concentrated to the inner ring area. In conclusion, a tailored ablation zone can be constructed by manipulating the resonant particle used in the ablation tool. Another important result is the thermal influence of the feeding microstrip line as shown in Fig. 6.50d. Due to its high thermal conductivity and its proximity to the SRR, it influences the hot spot creation dramatically and leads only to hot spots on the far side of the SRR independent of the SRR orientation and configuration. It therefore invalidates the freedom of hot spot creation by proper particle design. In order to exploit the shown possibilities of hot spot manipulation, a different feeding line structure with less thermal influence needs to be used. Promising structures seem to be CPW or slot line configurations because the SRRs can be effectively thermally isolated from the line by a dielectric layer between line and ring metallization. Ongoing work is being done with a CPW design where different resonant particles are being compared by placing them on the back side of the CPW line and the intermediate results look very promising.

References

1. State of Cancer Research. http://www.cancer.gov
2. P. Gascoyne, J. Noshari, F. Becker, R. Pethig, Use of dielectrophoretic collection spectra for characterizing differences between normal and cancerous cells. IEEE Trans. Ind. Appl. **30**(4), 829–834 (1994)
3. P. Gascoyne, X. Wang, Y. Huang, F. Becker, Dielectrophoretic separation of cancer cells from blood. IEEE Trans. Ind. Appl. **33**(3), 670–678 (1997)
4. P. Gascoyne, J. Vykoukal, Dielectrophoresis-based sample handling in general-purpose programmable diagnostic instruments. Proc. IEEE **92**(1), 22–42 (2004)

5. Y. Choongho, J. Vykoukal, D. Vykoukal, J. Schwartz, L. Shi, P. Gascoyne, A three-dimensional dielectrophoretic particle focusing channel for microcytometry applications. J. Microelectromech. Syst. **14**(3), 480–487 (2005)
6. H. Schwan, K. Foster, RF-field interactions with biological systems: electrical properties and biophysical mechanisms. Proc. IEEE **68**(1), 104–113 (1980)
7. H. Schwan, Analysis of dielectric data: experience gained with biological materials. IEEE Trans. Electr. Insul. **EI-20**(6), 913–922 (1985)
8. H.P. Schwan, Electrical properties of tissues and cell suspensions: mechanisms and models, in *16th Annual International Conference of the IEEE Engineering in Medicine and Biology Society*, Baltimore, Maryland, USA, IEEE, November 1994, pp. A70–A71. doi:10.1109/IEMBS.1994.412155
9. S. Gabriel, R.W. Lau, C. Gabriel, The dielectric properties of biological tissue: III. Parametric models for the dielectric spectrum of tissues. Phys. Med. Biol. **41**, 2271–2293 (1996)
10. D.A. Dean, D. Machado-Aranda, T. Ramanathan, I. Molina, R. Sundararajan, Electrical properties of biological tissues—an impedance spectroscopy study, in *IEEE Conference on Electrical Insulation and Dielectric Phenomena*, Kansas City, MO, USA, IEEE, October 2006, pp. 357–360. doi:10.1109/CEIDP.2006.311943
11. M.S. Boybay, O.M. Ramahi, Double negative metamaterials for subsurface detection, in *29th Annual International Conference of the IEEE Engineering in Medicine and Biology Society*, Lyon, France, IEEE, August 2007, pp. 3485–3488. doi:10.1109/IEMBS.2007.4353081
12. M. Schaefer, K. Nowak, B. Kherad, W. Gross, S. Post, M. Gebhard, Time domain reflectometry: measurement of free water in normal lung and pulmonary edema. Med. Biol. Eng. Comput. **42**(5), 577–580 (2004)
13. T.T. Thai, J.M. Mehdi, H. Aubert, P. Pons, G.R. DeJean, M.M. Tentzeris, R. Plana, A novel passive wireless ultrasensitive RF temperature transducer for remote sensing, in *IEEE MTT-S International Microwave Symposium Digest (MTT)*, Anaheim, USA, IEEE, May 2010, pp. 473–476. doi:10.1109/MWSYM.2010.5517892
14. Y. Xia, L. Wang, A wireless sensor using left-handed metamaterials, in *4th International Conference on Wireless Communications, Networking and Mobile Computing, 2008*, WiCOM '08, Dalian, China, IEEE, September 2008, pp. 1–3. doi:10.1109/WiCom.920
15. E. Ekmekci, C. Turhan-Sayan, Metamaterial sensor applications based on broadside-coupled SRR and V-shaped resonator structures, in *IEEE International Symposium on Antennas and Propagation (APSURSI)*, Washington, USA, IEEE, July 2011, pp. 1170–1172. doi:10.1109/APS.2011.5996492
16. M. Abidi, A. Elhawil, J. Stiens, R. Vounchx, J. B. Tahar, F. Choubani, Sensing liquid properties using split-ring resonator in Mm-wave band, in *IECON 2010—36th Annual Conference on IEEE Industrial Electronics Society*, Phoenix, USA, IEEE, November 2010, pp. 1298–1301. DOI:10.1109/IECON.2010.5675495
17. R. Melik, E. Unal, N. Perkgoz, C. Puttlitz, H. Demir, Metamaterial-based wireless strain sensors. Appl. Phys. Lett. **95**(1), 011106 (2009)
18. R. Melik, E. Unal, N. Perkgoz, B. Santoni, D. Kamstock, C. Puttlitz, H. Demir, Nested metamaterials for wireless strain sensing. IEEE J. Sel. Top. Quantum Electron. **16**(2), 450–458 (2010)
19. J.F. O'Hara, R. Singh, X.G. Peralta, I. Brener, E.A. Shaner, D.W. Branch, J. Han, A.J. Taylor, W. Zhang, Sensing liquid properties using split-ring resonator in Mm-wave band, in *Conference on Lasers and Electro-Optics, Quantum Electronics and Laser Science*, CLEO/QELS 2008, San Jose, USA, May 2008, pp. 1–2
20. P. Sprawls, *Physical Principles of Medical Imaging* (Lippincott Williams and Wilkins, Philadelphia, 1987)
21. J. Lin, *Electromagnetic Fields in Biological Systems* (CRC Press, Boca Raton, 2012)
22. C. Brace, Microwave ablation in *Tissue Ablation: Devices and Procedures* ed. by J.G. Webster (Wiley, Hoboken, 2005)
23. M. Puentes, C. Weiß, M. Schüßler, R. Jakoby, Sensor array based on split ring resonators for analysis of organic tissues, in *Proceedings of IEEE MTT-S International Microwave Symposium Digest*, (Baltimore, USA, June 2011)

24. J. Baena, J. Bonache, F. Martin, R. Sillero, F. Falcone, T. Lopetegi, M. Laso, J. Garcia-Garcia, I. Gil, M. Portillo, M. Sorolla, Equivalent-circuit models for split-ring resonators and complementary split-ring resonators coupled to planar transmission lines. IEEE Trans. Microwave Theory Tech. **53**(4), 1451–1461 (2005)
25. M. Puentes, M. Maasch, M. Schüßler, R. Jakoby, Frequency multiplexed 2-dimensional sensor array based on split-ring resonators for organic tissue analysis. IEEE Trans. Microwave Theory Tech. **60**(6), 1720–1727 (2012)
26. M. Puentes, M. Schüßler, R. Jakoby, 2D sensor array based on split rings resonators for monitoring of organic tissue, in *Proceedings of IEEE International Sensors Conference*, (Limerick, Ireland, October 2011)
27. A. Trehan, N. Nikolova, Numerical and physical models for microwave breast imaging. Department of Electrical and Computer Engineering, McMaster University, vol. 47, Issue 2 (2009)
28. Y. Okano, K. Ito, H. Kawai, Solid phantom composed of glycerin and its application to SAR estimation. IEICE Trans. Commun. **J83-B**(4), 534–543 (2000) (Japanese edition)
29. M. Puentes, F. Bashir, M. Maasch, M. Schüßler, R. Jakoby, Planar microwave sensor for thermal ablation of organic tissue, in *Proceedings of European Microwave Conference*, (Nuremberg, Germany, October 2013)
30. R. Habash, R. Bansal, D. Krewski, H. Alhafid, Thermal therapy, part III: ablation techniques. Crit. Rev. Biomed. Eng. **35**(1–2), 37–121 (2007)
31. C. Brace, Thermal tumor ablation in clinical use. IEEE Pulse **2**, 28–38 (2011)
32. M. Nikfarjam, V. Muralidharan, C. Christophi, Mechanisms of focal heat destruction of liver tumors. J. Surg. Res. **127**(2), 208–223 (2005)
33. M. Velazquez-Ahumada, M. Freire, R. Marques, Metamaterial applicator for microwave hyperthermia, in *Proceedings of URSI General Assembly and Scientific Symposium of International Union of Radio Science*, (Istanbul, Turkey, August 2011)
34. A. Strickland, P. Clegg, N. Cronin, B. Swift, M. Festing, K. West, G. Robertson, D.M. Lloyd, Experimental study of large-volume microwave ablation in the liver. Brit. J. Surg. **89**(8), 1003–1007 (2002)
35. F. Kreith, D. Goswami, *The CRC Handbook of Mechanical Engineering* (CRC Press, Boca Raton, 2005)
36. H. Pennes, Analysis of tissue and arterial blood temperatures in the resting forearm. J. Appl. Physiol. **1**, 93–122 (1948) (Republished 1998)
37. M. Puentes, F. Bashir, M. Schüßler, R. Jakoby, Dual mode microwave tool for dielectric analysis and thermal ablation treatment of organic tissue, in *Proceedings of 34th Annual International Conference of the Engineering in Medicine and Biology Society*, (San Diego, USA, August 2012)

Chapter 7
Conclusion and Outlook

The main goal of this dissertation was to study and developed new and innovative planar microwave sensor arrays. Sensors are used in every part of modern life and they are key components in many areas such as industrial, environmental, automotive, biomedical, among others. The reason for aiming the operation frequency of the sensors at the microwave range is due to the advantages that it can give to the sensing task. Microwave sensing enables the interaction between the sensor and the MUT in a non-invasive, non-ionizing and contact-less way. Furthermore, many information on the structure and function of many MUTs can be extracted at this frequency range.

In comparison to the common microwave sensors available on the market, an extra advantage is given to the sensors presented within the framework of this work by using the well-established theory of metamaterials. The use of periodic structures for the developed sensors allowed them to have the additional feature of spatial subwavelength resolution. They can give localized information within the sensor array of the relative dielectric changes and position by using a single structure and one readout signal. The principle of operation of all the presented sensors is based on the interaction between an electromagnetic field, protruding from the planar sensor surface, and the MUT located within the reach of this field. From the reflected and/or transmitted signal the information about the dielectric properties of the MUT and its location within the sensor array can be obtained through the extraction of a capacitive profile. The interesting part is that this principle can be customized for any application, where the dielectric properties of a material are of importance at this frequency range. The size, frequency and even the geometry offer a wide flexibility to adapt the principle to almost any environment. The advantages of using a periodic structure as the key component of a sensor were discussed in terms of different parameters that are interesting for sensing applications. Among this parameters are resonance sensitivity, phase sensitivity, delay sensitivity, resonator length, delay and finally harmonic bandwidth. It was stated that in every case the use of a metamaterial structure under certain design constraints could provide 4 to 10 times better performance depending on the parameter under consideration, as compared to a conventional RH design.

M. Puentes Vargas, *Planar Metamaterial Based Microwave Sensor Arrays for Biomedical Analysis and Treatment*, Springer Theses,
DOI: 10.1007/978-3-319-06041-5_7, © Springer International Publishing Switzerland 2014

In order to produce a relative dielectric image of the MUT, robust extraction mechanisms were developed. The simulations and measurements of the sensors normally produced only scattering parameters. S-parameters do not provide straightforward the dielectric information of the MUT. For this reason two different methods were studied and applied to the sensor structures. First, the multi-resonant perturbation method (MRPM) was directly taken from [1] and applied to the developed sensor devices. The method is able to extract the capacitive profile ($C_{LH,n}$) of a periodic structure influenced by a MUT. The second was completely developed within this thesis and it is based on network synthesis. It is called the mixed Cauer expansion method (MCEM) and it is able to extract the complete equivalent circuit of the periodic structure as long as it can be modeled as a CRLH-TL. Both methods proved to be comparable in terms of what they can achieve and are able to give a one unit cell resolution. The MCEM has the advantage of extracting the complete equivalent circuit of the structure without requiring any calibration measurement. Although it performed formidable for simulations, for measurement data the desired accuracy was not achieved due to the lossy materials that were analyzed specially for the biomedical applications. The main problem lies in the construction of an adequate impedance transfer function since the lossy material produces a shift in the imaginary part of the input impedance making it very difficult to extract its poles and zeroes. Also for large structures, the continued divisions of the polynomials inserted a high numerical error. As for the MRPM, it performed well both for simulation and measurements. The extracted profiles with this method are readable although slight errors of less than 10 % still occurred due to its linearization nature. The only disadvantage from this method is the tedious calibration measurements required, especially when the structures have a large number of unit cells. On the other hand, it has the great advantage of needing only information of one (RH or LH) band, since only one forth of the poles and zeros need to be detected, which can result in a cost reduction of the sensor.

The basic sensor concept was carefully developed and initial tests showed great promise in terms of sensitivity, cost, easy fabrication and great flexibility to be integrated in different sensing environments. The first test with a specific MUT was the monitoring of the quality of paper. For this purpose two different CRLH-TL based sensors were presented. The first developed planar sensor prototype consisted of a periodical arrangement of identical unit cells and each one includes a serial capacitor realized by two metallic patches and a shunt inductor realized with a thin long line between the patches and the ground. Since there is no ground plane on the backside of the substrate, it should be emphasized that it is not a microstrip circuit. The second design has also a periodic arrangement of unit cells and it is based on a CPW, where the capacitors are realized by two metallic patches and the inductors realized by gaps between the central strip of the CPW and the ground planes. It was proven that a material characterization with changes in the capacitance as small as 0.1 % can be done as long as the paper is located within 2.5 cm of the sensitive area of the sensor. Measurements with moist paper were successfully carried out, although this type of MUT is quite challenging for microwave sensors due to its highly lossy behavior.

To corroborate the high flexibility of the sensor concept to any sensing environment, two real applications were targeted. The first application is for monitoring biological liquids samples. The standard way to analyze pools of cells is done by looking at the sample under the microscope and using markers or fluorescent agents that might alter the biological activity of the MUT. Normally, a large amount of samples needs to be checked for particular properties such as concentration, protein function, the stage of the developing cycle, surface modification due to drug delivery, among others, and therefore the manual process takes a very long time. In this work, a microwave sensor array system made of planar periodic metamaterial structures combined with microfluidic channels technology was successfully developed. The presented prototypes enabled measurements of different concentration of water-ethanol mixtures. Furthermore, measurements of different concentrations of chinese hamster ovary (CHO) cells, a well-known mammalian cell model for biological and medical research, were successfully extracted using the MRPM with an error smaller than 0.1 %.

The second and final area of application to be explored was for the medical field. Since organic tissues have a frequency dependant dielectric behavior that can be monitored to extract information such as the presence of abnormalities, it is logical to explore different sensing mechanisms at different operating frequencies for that purpose. Microwave techniques offer an advantage over classical methods such as computerized tomography (CT) scans, X-Rays or magnetic resonance imaging (MRI), since they can provide cost-effective devices that can localize with high precision an abnormality within the human body and may even present treatment options for certain types of cancer in a minimal invasive way. In this work, different microwave planar sensors based on SRRs were presented with optimized designs to perform the sensing of the dielectric properties of different organic tissues. Furthermore, the same structures were designed to offer a second mode of operation and perform treatment of cancerous tissues by means of thermal ablation. The measurements were carried out using adapted phantoms of organic tissue and the presented prototypes were able to heat a considerable ablation zone to approximately 109 °C with an input power of 10 W. This approach of sensing and treatment by thermal ablation with the same device offers several benefits such as, the automatic alignment between the determined tumor location and subsequent heat application. Furthermore, a control of the ablated area can be done by using the tool in the sensing mode after the ablation procedure is completed.

In summary, a new innovative sensor concept for planar microwave sensor arrays was presented. It was proven that the use of metamaterial structures is very useful since they can not only increase the performance of the sensor devices but combined with appropriate extraction techniques, they can give a frequency characterization and the spatial distribution of the MUT. They offer incredible flexibility in terms of operating frequency and geometry, making them good candidates for many sensor applications. Therefore, the novelty of the work presented in this thesis is definitively the combination of periodic structures, one of them with an interesting new geometry, with robust extraction techniques to create sensor arrays that can monitor several samples of MUT simultaneously in a semi automated way, and furthermore, interact with the MUT when needed to create physical changes as shown with the thermal ablation sensors.

7.1 Outlook

The future steps for the presented work will be derived by its application areas. In the biological application, first a dedicated setup for the precise and automated control in the filling of the channel is a priority. With this setup it is possible to increase the amount of unit cells in the sensor. The new design should aim for a lab-on-chip system and the chosen applicative case to continue this work, within the framework of a joint project with the CNRS-LAAS institute, is the study of chronic lymphocytic leukemia and the cell response to various biochemical drugs. This application is of high interest since chronic lymphocytic leukemia is the most common leukemia in the western hemisphere, with a heterogeneous 5 years relative survival expectancy.

For the medical application, work will continue on the topic of thermal ablation. The microwave devices have proven themselves as worthy competitors to the RF ablation devices and the array feature gives them an extra advantage as well as the required low input power. Future work will be oriented in trying to make the transfer of the acquired knowledge into an actual device that could undergo clinical trials. The initial step should be to make measurements with animal subjects to test the devices under a more real scenario and make the appropriate adjustments to the design. On the other hand, heat can be used not only to destroy cells, but also to aid in the recovery of cells and tissues. Such use of heat is sometimes called therapeutic hyperthermia to distinguish it from malignant hyperthermia. The application of heat to injured tissues can help in the recovery through a temperature dependent vasodilation. In general, vasodilation increases the mass transport of wastes and nutrients from and to the site of an injury, thus by applying heat the damaged tissue is metabolically more active and the mass transport capabilities are enhanced facilitating a more rapid healing. Therefore, a completely new field of application for the presented devices is open for future exploration.

Reference

1. M. Schüßler, M. Puentes, C. Mandel, R. Jakoby, Multi-Resonant Perturbation Method for Capacitive Sensing with Composite Right-Left Handed Transmission Line, in *Proceedings of IEEE MTT-S International Microwave Symposium Digest*, (Anaheim, USA, June 2010)

Appendix A

A.1 Phantoms of Organic Tissue

The use of phantoms that mimic the dielectric properties of tissue can help create accurate models for initial testing of devices before going into human trial phase. The phantom must be able to mimic the electric field distribution and relative heating patterns of the actual tissue in the presence of an electromagnetic source. The recipes for the developed phantoms used in this work were taken from [1] and [2]. A short description of the fabrication process will be addressed. Each phantom recipe simulates a single type of tissue. It exhibits a homogenous distribution of electrical properties specific to the target tissue. Nevertheless, the human body tissues are normally heterogenic and a layering technique in the fabrication of the phantoms can address this difficulty. The materials use for the development of phantoms are non biological which gives them a longer shelf life and are less prone to degradation than real tissues. Additionally, they mimic the properties of human tissue more accurate than animal tissues.

The desired properties of the phantoms are mechanical stability of shape, non-toxicity, fabrication simplicity, wideband properties, ease maintenance and low cost. The most typical phantoms found in the literature are the glycerin phantom (first introduced by Yoshiki Okano) and the TX-151 phantom (first introduced by Koichi Ito) [3]. The glycerin phantoms are chosen since they comply with all the requirement in contrary to the TX-151 phantoms. The main ingredients in the glycerin phantoms are glycerin, water, salt, agar and polyethylene powder (PEP). Agar acts as a coagulant and gives mechanical strength to the phantom, water controls the permittivity, salt the conductivity and PEP is used for fine tuning of both dielectric parameters. Glycerin phantoms can mimic both high, intermediate and low-water content tissues by varying the percentage of water, PEP and salt. Nevertheless, above 3 GHz the losses can be controlled by water and PEP alone and healthy tissue phantoms do not require salt.

In the case of malignant tissue phantoms, the permittivity and conductivity are higher than for normal tissues. Hence, water and salt (NaCl) are the main constituents of most tumor recipes. PEP cannot be used as the base powder as it lowers the

M. Puentes Vargas, *Planar Metamaterial Based Microwave Sensor Arrays for Biomedical Analysis and Treatment*, Springer Theses,
DOI: 10.1007/978-3-319-06041-5,

Table A.1 Recipe for high water content tissue phantom (Muscle)

Materials	Weight % based on glycerine
Glycerine	100
DIW	180
PEP	15
Agar	12

Table A.2 Recipe for high water content tissue phantom (FG tissue)

Materials	Weight % based on glycerine
Glycerine	100
DIW	140
PEP	14
Agar	11

Table A.3 Recipe for intermediate water content tissue phantom (Skin and transitional tissue)

Materials	Weight % based on glycerine
Glycerine	100
DIW	100
PEP	8
Agar	8

Table A.4 Recipe for low water content tissue phantom (Adipose fat)

Materials	Weight % based on glycerine
Glycerine	100
DIW	8
PEP	10
Agar	11.5
Ethylene glycol	52

dielectric properties. Alginate powder or gelatine are the most common alternatives to make tumors.

The ingredients for the recipe of a tissue with very high water content such as muscle is given in Table A.1. It should be mentioned that the ingredients of all the phantom recipes have glycerin as the percentage base value, i.e., the value of all other materials are given as percentage weight calculated with respect to glycerin. The recipe for another type of high water content tissue phantom, very important for breast phantom models, is for fibro-glandular (FG) tissue and is given in Table A.2. For intermediate water content tissue phantom, such as skin or transitional tissue, the values are given in Table A.3. The final recipe is for low water content tissue phantom, such as adipose fat, and is depicted in Table A.4.

Table A.5 Recipe for malignant tissue based on alginate powder

Materials	Quantity (g)
Alginate powder	18
DIW	100
NaCl	0.7

Table A.6 Recipe for malignant tissue based on gelatine

Materials	Quantity (g)
Gelatine	7
DIW	100
NaCl	0.7

In the case of malignant tissue, i.e., tumors; there are two different recipes given in Tables A.5 and A.6. For these receipts no glycerin is used and the given values are specific quantities. All phantoms can be preserved for six months by wrapping it in vinyl/plastic film and storing it in refrigeration.

For the fabrication procedure, a manual methodology is used as long as the maximum value of the phantom mixture is below 800 ml. If is above this value is recommended the use of an automatic kitchen food processor for blending the ingredients. For the phantoms presented in this work the manual methodology was sufficient. The steps for the fabrication of the glycerin based phantoms are [1]:

- Wear gloves, lab goggles and a lab coat. Follow standard laboratory safety regulations.
- Pour DIW into a suitable glass flask (narrow opening of flask is preferred to a cylindrical beaker so as to restrict air bubbles) and put it on a low flame on a burner. A mesh is recommended over the burner to avoid direct contact of the flask with flame. It is essential for better control of temperature.
- Pour all liquids relevant to recipe and stir to form a clear solution. Heat the solution to approximately 46 °C.
- Add bactericide (a few drops) and stir it to a clear solution.
- Add the powder reagents, which are the mixture of PEP and agar for these recipes. The powder mixture is added in tiny amounts to the liquid mixture. The mixture is stirred continuously, gently and carefully to form a homogenous solution with as less bubbles as possible. Near the solidification point the bubbles escape. Do not stir very briskly or hard as this weakens polymerization reaction and phantoms are susceptible to easy breakdown post fabrication.
- Once the mixture is homogenous and thick, pour it in a suitable mould by holding the hot flask with a tong or cloth. Usually the temperature reached is in the range 68–75 °C.
- After initial cooling, keep the phantom in the refrigerator for solidification via faster cooling.

- Cover the mould with vinyl/plastic film once the phantom solidifies. Also use high quality plastic wraps (such as Para-film) to store the phantom post fabrication for longer use.

In the case of malignant tissue phantom the fabrication steps vary and are the following:

- Wear gloves, lab goggles and a lab coat. Follow standard laboratory safety regulations.
- Pour water in a flask and heat it to a warm temperature.
- Dissolve salt in it and make a clear solution.
- Add a few drops of bactericide and put off the flame.
- For gelatine-based recipe, dissolve the gelatine powder directly in warm saline solution and set it to solidify. Whereas, for alginate powder, first let the saline cool down to room temperature and then dissolve the alginate powder quickly in saline to form a thick mixture. The alginate powder solidifies within 10 seconds. Hence immediately scoop out the alginate-saline mixture in specific mould. If this is not done in time, the only way to use the phantom is by cutting specific shapes from the solidified phantom chunk. This is not a big issue considering that irregular and random shapes are present in malignant tumors.

References

1. A. Trehan, N. Nikolova, *Numerical and Physical Models for Microwave Breast Imaging*, Department of Electrical and Computer Engineering, McMaster University, vol. 47, Issue 2, 2009
2. A. Trehan, N. Nikolova, *Summary of Materials and Recipes available in the Literature to Fabricate Biological Phantoms for RF and Microwave Experiments*, Computational Electromagnetics Laboratory, Department of Electrical and Computer Engineering, McMaster University, vol. 47, Issue 2, 2009
3. Y. Okano, K. Ito, H. Kawai, Solid phantom composed of glycerin and its application to SAR estimation. IEICE Trans. Commun. **J83–B**(4), 534–543 (2000) (Japanese Edition)

About the Author

Dr.-Ing Margarita Puentes Vargas was born in Bucaramanga, Colombia in 1981. She graduated with honors as Electronic Engineer in 2001 from the University El Bosque in Bogota, Colombia. She graduated as Master in Electronic Engineering and Computers in 2006 from the University Los Andes in Bogota, Colombia. In 2013 she receive her Ph.D. degree with "Summa Cum Laude" from the Technische Universität Darmstadt, Germany. Her current research is focused on new and innovative metamaterials applications especially for biomedical sensors. She won the Best Paper Award from the German Microwave Conference in 2009 and as a co-author received as well the Best Graduate Student Paper Award from the IEEE International Instrumentation and Measurement Technology Conference in 2009.

M. Puentes Vargas, *Planar Metamaterial Based Microwave Sensor Arrays for Biomedical Analysis and Treatment*, Springer Theses,
DOI: 10.1007/978-3-319-06041-5, © Springer International Publishing Switzerland 2014

MIX
Papier aus verantwortungsvollen Quellen
Paper from responsible sources
FSC® C105338

If you have any concerns about our products,
you can contact us on
ProductSafety@springernature.com

In case Publisher is established outside the EU,
the EU authorized representative is:
Springer Nature Customer Service Center GmbH
Europaplatz 3, 69115 Heidelberg, Germany

Printed by Libri Plureos GmbH
in Hamburg, Germany